Principles and Standards for School Mathematics Navigations Series

Navigating *through* Geometry *in* Grades 6–8

David K. Pugalee
Jeffrey Frykholm
Art Johnson
Hannah Slovin
Carol Malloy
Ron Preston

Susan N. Friel
Grades 6–8 Editor
Peggy A. House
Navigations Series Editor

NATIONAL COUNCIL OF TEACHERS OF MATHEMATICS

1906 Association Drive, Reston, VA 20191-9988
(703) 620-9840; (800) 235-7566
www.nctm.org

ISBN 0-87353-513-8

Printed in the United States of America

Table of Contents

Appendix

Blackline Masters and Solutions 85

Contents of CD-ROM

Readings and Supplemental Materials

Essay

About This Book

> When we see how the branching of trees resembles the branching of arteries and the branching of rivers, how crystal grains look like soap bubbles and the plates of a tortoise's shell, how the fiddleheads of ferns, stellar galaxies, and water emptying from the bathtub spiral in a similar manner, then we cannot help but wonder why nature uses only a few kindred forms in so many different contexts. Why do meandering snakes, meandering rivers, and loops of string adopt the same pattern, and why do cracks in mud and markings on a giraffe arrange themselves like films in a froth of bubbles?
>
> —Peter S. Stevens, *Patterns in Nature*

Have you ever wondered why bees store their honey in hexagonal rather than octagonal honeycombs? Or why a three-legged stool doesn't wobble? Have you considered why you need a mirror that is at least half your height in order to see your entire body? Indeed, from the alignment of the solar system to the structure of an atom, from rocks to crystals to flowers to rings on a snake, from architects to mechanics to artists to musicians, from bike gears to curve balls to Snowboards and Rollerblades, geometry pervades our world.

The study of geometry in the middle grades is crucial in the mathematics education of our children. During these years, students begin to develop the cognitive structures that allow them to reason within a linear, deductive system of thought. Simultaneously, they continue informally to observe the boundless activity of their environment, tapping into native curiosities and intuitions that can, if nurtured, provide the foundation for inductive discoveries and reinventions of many of our fundamental mathematical concepts. Teaching geometry in the middle grades requires delicate pedagogical navigation: How does a teacher draw students toward more-sophisticated, formal systems of thought while recognizing their still wide ranging cognitive abilities and experiences? How does a teacher facilitate the transition between hands-on learning experiences and more-formal, abstract approaches to learning geometric concepts? These questions and many others like them will be addressed through the activities in this book.

As learners progress through the middle-grades curriculum toward a more formal study of geometry, it is important that they begin making the transition from empirically based, inductive reasoning to the deductive use of rules and abstract thinking that are the hallmarks of more-advanced geometry. As *Principles and Standards for School Mathematics* (National Council of Teachers of Mathematics [NCTM] 2000) suggests, students in the middle grades must begin to "create and critique inductive and deductive arguments concerning geometric ideas and relationships" (p. 232). The development of mathematical arguments in the middle grades promotes the transition from informal to more-formal thinking, which leads to an emphasis on mathematical reasoning, including inductive and deductive processes, formulating and defending conjectures, and classifying and defining geometric objects. This book focuses on the central concepts of middle-grades geometry and students' geometric thinking.

The van Hiele framework can provide insights into the development of cognitive and spatial processes in middle-grades students. The van Hieles described five levels of thought that characterize the development of the reasoning abilities needed to engage in geometric thinking. In an essay on the CD-ROM that accompanies this book, Carol Malloy supplies an introduction to these levels and a discussion of their relevance to designing instruction for middle-grades students. Related readings also appear on the CD-ROM.

The four chapters in this book emphasize geometric thinking as an expected outcome of the mathematics experiences of students in grades 6–8. The chapters present important ideas and relevant activities that focus on the "big ideas" of geometry—shape, location, transformations, and visualization. Paralleling the Geometry Standard, the chapters emphasize the expectations for middle-grades students in each of these areas:

- Chapter 1, "Characteristics and Properties of Shapes," emphasizes the development of geometric reasoning.
- Chapter 2, "Coordinate Geometry and Other Representational Systems," explores the use of multiple representations as tools for analysis.
- Chapter 3, "Transformations and Symmetry," emphasizes the use of transformation geometry as another lens for investigating and interpreting geometric objects.
- Chapter 4, "Visualization, Spatial Reasoning, and Geometric Modeling," emphasizes the development of reasoning and the ability to visualize relationships.

Each section of a chapter presents one or more sample activities, many of which have blackline masters, which are signaled by an icon and can be found in the appendix, along with solutions to the problems where appropriate. They can also be printed from the CD-ROM that accompanies the book. The CD-ROM, also signaled by an icon, contains four applets for students to manipulate and resources for professional development. An icon also appears in the margin next to references to *Principles and Standards for School Mathematics.*

The authors gratefully acknowledge the contributions of the following people, who gave us the benefit of their experience as classroom teachers:

Sandra W. Childrey
Neal Kaufmann
Donna Parrish

Key to Icons

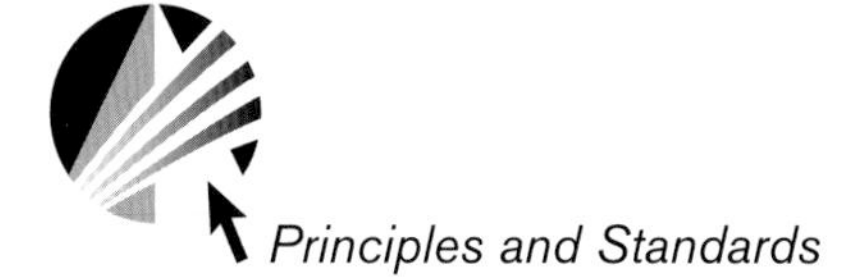

Three different icons appear in the book, as shown in the key. One alerts readers to material quoted from *Principles and Standards for School Mathematics,* another points them to supplementary materials on the CD-ROM that accompanies the book, and a third signals the blackline masters and indicates their locations in the appendix.

GRADES 6–8

NAVIGATING *through* GEOMETRY

Introduction

Both in the development of mathematics by ancient civilizations and in the intellectual development of individual human beings, the spatial and geometric properties of the physical environment are among the first mathematical ideas to emerge. Geometry enables us to describe, analyze, and understand our physical world, so there is little wonder that it holds a central place in mathematics or that it should be a focus throughout the school mathematics curriculum.

When very young children begin school, they already possess many rudimentary concepts of shape and space that form the foundation for the geometric knowledge and spatial reasoning that should develop throughout the years. *Principles and Standards for School Mathematics* (National Council of Teachers of Mathematics [NCTM] 2000) recognizes the importance of a strong focus on geometry throughout the entire prekindergarten–grade 12 curriculum, a focus that emphasizes learning to—

- analyze characteristics and properties of two- and three-dimensional geometric shapes and develop mathematical arguments about geometric relationships;
- specify locations and describe spatial relationships using coordinate geometry and other representational systems;
- apply transformations and use symmetry to analyze mathematical situations;
- use visualization, spatial reasoning, and geometric modeling to solve problems. (P. 41)

Geometry not only provides a means for describing, analyzing, and understanding structures in the world around us but also introduces an experience of mathematics that complements and supports the study of other aspects of mathematics such as number and measurement. Geometry offers powerful tools for representing and solving problems in all areas of mathematics, in other school subjects, and in everyday applications. *Principles and Standards* presents a vision of how geometric concepts and reasoning should develop and deepen over the course of the school mathematics curriculum. The *Navigating through Geometry* books elaborate that vision by showing how important geometric concepts can be introduced, how they grow, what to expect of students during and at the end of each grade band, how to assess what students know, and how representative instructional activities can help translate the vision of *Principles and Standards* into classroom practice and student learning.

Foundational Components of Geometric Thinking

The Geometry Standard emphasizes as major unifying ideas *shape* and the ability to analyze characteristics and properties of two- and three-dimensional objects and develop mathematical arguments about geometric relationships; *location* and the ability to specify positions and describe spatial relationships using various representational systems; *transformations* and the ability to apply motions, symmetry, and scaling to analyze mathematical situations; and *visualization* and the ability to create and manipulate mental images and apply spatial reasoning and geometric modeling to solve problems. Each of these components of geometric thinking requires nurturing and developing throughout the school curriculum.

Analyzing characteristics and properties of shapes

By the time the youngest children begin formal schooling, they have already formed many concepts of shape, although their understanding is largely at the level of recognizing shapes by their general appearance and they frequently describe shapes in terms of familiar objects such as a box or a ball. In the primary grades, children should have ample opportunities to refine and focus their understanding and to gradually develop a mathematical vocabulary. They also should learn to recognize and name the parts of two- and three-dimensional shapes, such as the sides and the "corners," or vertices. Teachers should provide frequent hands-on experiences with materials, including technology, that help the students focus on attributes of various shapes, such as that a square is a special rectangle with all four sides the same length or that pyramids always have triangular faces that meet at a common point. Experiences that promote such outcomes include building and drawing shapes; comparing shapes and describing how they are alike and how they are different; sorting shapes according to one or more attributes; cutting or separating shapes into component parts and reassembling the parts to form the original or different shapes; and identifying shapes found in everyday objects or in the classroom, home, or neighborhood. Throughout such activities, teachers must take care to ensure that the children encounter both examples and nonexamples of common shapes and that they see those examples in many different contexts and orientations so that they learn to identify a triangle or a rectangle, for example, no matter what material it is made of or how it is positioned in space.

As children progress to the higher elementary grades, they should continue to identify, compare, classify, and analyze increasingly more complex two- and three-dimensional shapes, and they should expand their mathematical vocabulary and refine their ability to describe shapes and their attributes. As they do so, they begin to develop generalizations about classes of shapes, such as prisms or parallelograms, and to formulate definitions for those classes. They also include in their study not only two- and three-dimensional shapes but points, lines, angles, and more-precise relationships such as parallelism and perpendicularity. They begin to explore properties of area and perimeter and to pose questions related to those measurement concepts; they might, for example, use tangram pieces to investigate whether shapes that are different, such as a rectangle, a trapezoid, and a nonrectangular parallelogram, can have the same area. They also develop and explore concepts of congruence and similarity, which they express in terms of shapes that "match exactly" (congruence) or shapes that "look alike" except for "magnifying" or "shrinking" (similarity). In grades 3–5, there should be a growing emphasis on making conjectures about geometric properties and relationships and formulating mathematical arguments to substantiate or refute those conjectures; for example, students might use tiles or grid paper to show that whenever the sides of one square are twice as long as the sides of another square, then four of the smaller squares will "fit inside" or "cover" the larger square, or they might measure to demonstrate that a rectangle, trapezoid, and nonrectangular parallelogram that have equal area do not necessarily have the same perimeter.

The informal knowledge and intuitive notions developed in the elementary grades receive more-careful examination and more-precise description in the middle grades. Descriptions, definitions, and classification schemes take account of multiple properties, such as lengths, perimeters, areas, volumes, and angle measures, and students should use those characteristics to analyze more-sophisticated relationships by, for instance, developing a classification scheme for quadrilaterals that accurately represents some classes of quadrilaterals (e.g., squares) as special cases or subsets of other classes (e.g., rectangles or rhombuses). At the same time, they should develop the more precise language needed to communicate ideas such as that all squares are rectangles but not all rectangles are squares. Students in the middle grades should also investigate what properties of certain shapes are necessary and adequate to define the class; they might explore, for example, the following question: Among the many characteristics of rhombuses, including congruent sides, opposite sides parallel, opposite angles congruent, diagonals that bisect each other, and perpendicular diagonals, which characteristics can be used to define rhombuses and to differentiate them from all other quadrilaterals? In a similar manner, other concepts introduced informally in the lower grades, including *congruence* and *similarity*, should be established more precisely and quantitatively during the middle grades, and special geometric relationships, including the Pythagorean relationship and formulas for determining the perimeter, area, and volume of various shapes, should be developed and applied. All these explorations should be carried out with the aid of hands-on materials and Dynamic Geometry® software, and all should be conducted in an environment in which students are expected and encouraged to make and test conjectures and develop convincing arguments, based on both inductive and deductive reasoning, to justify their conclusions.

By the time students reach high school, they should be able to extend and apply the geometric knowledge developed earlier to establish or refute conjectures, deduce new knowledge from previously established facts, and solve geometric problems. They should be helped to extend the knowledge gained from specific problems or cases to more-general classes of objects and thus to establish the validity of geometric conjectures, prove theorems, and critique arguments proposed by others. As they do so, students should organize their knowledge systematically in order to understand the role of definitions and axioms and to appreciate the connectedness of logical chains, recognizing, for example, that if a result is proved true for an arbitrary parallelogram, then it automatically applies to all rectangles and rhombuses.

Specifying locations and describing spatial relationships

The importance of location and spatial relationships becomes apparent when we try to answer questions such as Where is it? (location), How far is it? (distance), Which way is it? (direction), and How is it oriented? (position). Typically, the first answers that children give to questions such as these are in relation to other objects: on the chair, next to the book, under the bed. In the primary grades, teachers help students develop a sense of location and spatial relationship by developing those early ideas. Using physical objects, often to illustrate stories, or physically acting out a relationship, children learn the meaning of such concepts as above, below, in front, behind, between, to the left, to the right, next to, and other relative positions. In time they add concepts of distance and direction, such as three steps forward, and they learn to combine such descriptions to lay out routes (e.g., walk to the door, turn left, go to the end of the corridor). Students begin to represent such physical notions of location, distance, and space both as verbal instructions and as diagrams or maps, and they learn to follow verbal directives and to read maps as means to locating a hidden object or reaching a desired destination. As their skills in representing locations increase, students should add more quantitative details by, for instance, pacing off or measuring distances to better communicate "how far" or adding a simple coordinate system to define a location more precisely.

In grades 3–5, students' understanding of location, direction, and distance are applied to increasingly more complex situations. They become more precise in their measurements and begin to examine situations to determine whether there is more than one route between two points or if there is a shortest distance between them. During these years, students should come to recognize that some positional representations are relative (e.g., *left* or *right*) or subjective (e.g., *near* or *far*), whereas others are fixed (e.g., *north* or *west*) and unambiguous (e.g., *between*) and that directions are not always interchangeable (e.g., two blocks north, then three blocks west does not take you to the same destination as two blocks west, then three blocks north, but three blocks west, then two blocks north does have the same end point as two blocks north, then three blocks west). They also should become more attentive to orientation; they might determine the direction that an object faces, whether it has been reflected or rotated from its initial position, or the distance and direction that it has been moved, all of which are closely related to ideas of transformations discussed later. It is particularly convenient and appropriate for students to explore the concepts of location and position by using grids together with graphical representations, physical models, and computer programs; as they continue their explorations on a grid, students also should learn to specify ordered pairs of numbers to represent coordinates and to use coordinates in locating points, describing paths, and determining distances along grid lines.

In the middle grades, the ideas established in elementary school should continue to be developed, and in addition, geometric ideas of location and distance can be linked to developing algebraic concepts as students apply coordinate geometry to the study of shapes and relationships. For example, the study of linear functions in algebra is related to the determination of the slopes of the line segments that form the sides of polygons, and these values in turn are used to determine relationships such as parallelism or perpendicularity of sides, which are used to analyze and classify the polygons; the Pythagorean relationship is applied to the coordinate plane to establish a method of determining the distance between points or the lengths of segments; and coordinates can be used to locate the midpoints of segments.

High school students should extend the geometric concepts of Cartesian coordinates used in lower grades to other coordinate systems, including polar, spherical, or navigational systems, and use them to analyze geometric situations. They also should develop facility in translating between different coordinate representations and should understand that each representation offers certain advantages in specific situations. During the secondary school years, students also learn to apply trigonometric relationships to solve problems involving location, distance, direction, and position, and analytical methods continue to be used, further strengthening the connection between algebra and geometry.

Applying transformations and symmetry

Among the early geometric discoveries that children make is that shapes can be moved without being changed: a triangle is still the same triangle even if it is flipped over or slid across the table, and a puzzle piece may need to be turned in order for it to fit into the desired space. Such intuitions are the starting point for studying transformations when children enter school. This important aspect of spatial learning in the primary school years engages students in exploring the motions of slides, flips, and turns, which leads to the discovery that such motions alter an object's location or orientation but not its size or shape. Primary-grades teachers should guide children to look for, describe, and explore symmetric shapes, which they can do informally by folding paper, tracing, creating designs with tiles, and investigating reflections in mirrors. Explorations that children enjoy in the primary grades, such as folding the net of a rectangular prism to make a "jacket" for a block, also involve relationships between two- and three-dimensional shapes.

As children move into the upper elementary grades, the more-informal notions of slides, flips, and turns are treated with greater precision as translations, reflections, and rotations, and attention is directed to what parameters must be specified in order to describe those transformations (e.g., slide [translate] the square ten centimeters to the right; flip [reflect] the triangle over its hypotenuse; turn [rotate] the drawing a quarter-turn clockwise). Students also learn that transformations can be used to demonstrate that two shapes are congruent if one can be moved so that it exactly coincides with the other. They should then be helped to extend that notion by being challenged to visualize and mentally manipulate shapes, describing mathematically a series of motions that can be used to demonstrate congruence or predicting the result of certain transformations before actually performing them with physical objects or symbolic representations. Such growing precision extends as well to symmetry as students learn to specify all the reflection lines or the center and the degrees of rotation in a symmetric figure or design.

In grades 6–8, transformation geometry can be a powerful tool for exploring spatial and geometric ideas. Not only are rigid transformations used to deepen students' understanding of such concepts as congruence, symmetry, and the properties of polygons, but dilations and the notions of scaling and similarity, which are closely linked to proportional reasoning, are introduced in the middle grades. Additionally, in their study of transformations, middle-grades students should, for instance, generalize the result of two successive reflections over parallel lines and compare that outcome to the result of two successive reflections over intersecting lines. They should also begin to quantify and formalize aspects of transformations, establishing, for example, that in a reflection each point on the original object is the same distance from the mirror line as the corresponding point on the image. Physical manipulatives, such as mirrors or other reflective devices, and Dynamic Geometry software are especially useful in conducting such investigations.

In high school, the study of transformations can be further enriched by the use of function notation, coordinates, vectors, and matrices to describe and investigate transformations, including both isometries and dilations. Students should develop certain basic "tools" such as determining a matrix representation for accomplishing a reflection over the line $y = x$ or other common transformations, and they should relate the composition of transformations to matrix multiplication and apply those concepts to the solution of problems.

Using visualization, spatial reasoning, and geometric modeling

The ability to create mental images of two- and three-dimensional objects, to visualize how objects appear from different perspectives, to formulate representations of how objects are positioned relative to other objects, to relate two-dimensional renderings to the three-dimensional objects that they represent, to predict how appearances will vary as the result of one or more transformations, and to create spatial representations to model various mathematical situations are among the most important outcomes of the study of geometry.

Young children begin to develop their spatial visualization by initially manipulating physical objects and later extending their manipulations to mental images. Teachers in the primary grades may help children develop spatial memory and spatial visualization by asking them to recall and describe hidden objects or by having them describe how an object would look if viewed from a different side. They may ask children to imagine, and later explore and verify, what will happen when a given shape is cut in two in a certain way or to predict and demonstrate what other shapes could result if that same object were cut in a different manner. Children should also experiment with different shapes and formulate descriptions of them, perhaps by creating a shape from tangrams and taking turns to describe what each one sees in the figure. Students should also learn to read and draw simple maps and to give and follow directions—for example, giving a classmate verbal instructions for going from the classroom to the cafeteria. Opportunities abound to develop spatial visualization in connection with other topics and subjects by, for instance, demonstrating that even numbers can always—whereas odd numbers can never—be arranged in two equal rows or highlighting spatial concepts during art or physical education lessons. Children should have ample opportunity to discover that spatial reasoning and geometric modeling can contribute to understanding and solving a wide variety of problems that involve number, data, and measurement and that have numerous applications.

As students move through grades 3–5, they become more adept at reasoning about spatial properties and relationships among shapes; they might develop strategies to calculate the area of a garden plot by subdividing it into component rectangles or relate the area of a trapezoid or a parallelogram to the area of the rectangle that is formed by cutting and reassembling the original quadrilateral. Relating three-dimensional shapes to their two-dimensional representations becomes an important topic in these grades as students discover how to build three-dimensional objects from two-dimensional representations, and vice-versa; construct and fold nets of solids; examine diagrams of nets to predict which ones can or cannot be folded to form a certain prism; or mentally manipulate a shape to produce an accurate picture of hidden parts. Applying geometric reasoning and modeling to solve problems in all areas of mathematics, as well as in other contexts, should continue to be a principal focus of the curriculum.

The skills of spatial visualization and geometric reasoning that emerge in the lower grades should become more refined and sophisticated in grades 6–8 as students solve problems involving distance, area, volume, surface area, angle measure, and other quantifiable properties. Students should be guided to develop, understand, and apply important formulas for calculating the length, area, or volume of selected shapes. As they explore relationships using physical models and appropriate technology, students should begin to establish and give arguments to support important generalizations; they might, for example, demonstrate why, when the side of a cube is tripled, the surface area of the enlarged cube is nine times the surface area of the original whereas the volume of the enlargement is twenty-seven times that of the original. Geometric models for algebraic and numerical relationships help students integrate important concepts from all strands of the mathematics curriculum; manipulatives and computer programs that connect geometric, algebraic, and numerical concepts contribute to students' developing mathematical maturity and enable them to solve more-complex problems both within mathematics and in other subjects.

As students progress through high school, their visualization skills should extend from representations on the familiar two- and three-dimensional rectangular coordinate systems to analogous representations on a spherical surface or in a spherical space; investigations that connected two-dimensional representations of polyhedra with three-dimensional representations of them later evolve into challenges of projecting a spherical surface onto a plane and producing a two-dimensional map of a three-dimensional surface. Producing perspective drawings, visualizing the resultant cross section when a plane slices a solid object, predicting the three-dimensional shape that results when a plane figure is swept 360 degrees about an axis, and navigating in a spherical frame of reference are examples of spatial ideas that should evolve as students progress through school. In high school, too, geometric representations can be of great benefit when studying topics involving algebra, measurement, number, and data, and the application of geometric ideas to the solution of problems across mathematics and in other disciplines is one of the major goals of the curriculum.

Developing a Geometry Curriculum

A curriculum that fosters the development of geometric thinking envisioned in *Principles and Standards* must be coherent, developmental, focused, and well articulated, not simply a collection of lessons or activities. Geometric ideas should be introduced in the earliest years of schooling and then must deepen and expand as students progress through the grades. As they move through school, children should receive instruction that links to, and builds on, the foundation of earlier years; they must continually be challenged to apply increasingly more sophisticated spatial thinking to solve problems in all areas of mathematics as well as in other school, home, and life situations.

These Navigations books do not attempt to describe a complete geometry curriculum. Rather, the four *Navigating through Geometry* books illustrate how selected "big ideas" of geometry develop across the prekindergarten–grade 12 curriculum. Many of the concepts presented in these geometry books will be encountered again in other contexts related to the Algebra, Number, Measurement, and Data Standards; in the Navigations books, as in the classroom, the concepts described under the Geometry Standard reinforce and enhance students' understanding of the other strands.

Geometry is essential to the vision of mathematics education set forth in *Principles and Standards for School Mathematics* because the methods and ideas of geometry are indispensable components of mathematical literacy. The *Navigating through Geometry* books are offered as guides to help educators set a course for successful implementation of the very important Geometry Standard.

GRADES 6–8

NAVIGATING *through* GEOMETRY

Chapter 1
Characteristics and Properties of Shapes

Important Mathematical Ideas

As students analyze two- and three-dimensional shapes by exploring their characteristics and properties, they should become more adept at developing mathematical arguments about geometric relationships. Specifically, it is important for students in grades 6–8 to have opportunities to—

- precisely describe, classify, and understand relationships among types of two- and three-dimensional objects using their defining properties;
- understand relationships among angles, side lengths, perimeters, areas, and volumes of similar objects;
- create and critique inductive and deductive arguments concerning geometric ideas and relationships, such as concurrence, similarity, and the Pythagorean relationship. (National Council of Teachers of Mathematics [NCTM], p. 232)

Figures 1.1 and 1.2 display classifications of two- and three-dimensional shapes.

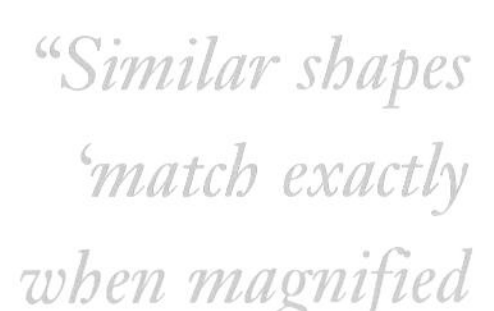

"Similar shapes 'match exactly when magnified or shrunk' … their corresponding angles are congruent and their corresponding sides are related by a scale factor." (NCTM 2000, p. 234)

What Might Students Already Know about These Ideas?

As articulated in *Principles and Standards for School Mathematics*, students in grades 3–5 should have opportunities to develop clarity and precision in describing properties of geometric shapes. Knowledge about how geometric shapes are related to one another is a foundation on which students can begin to build geometric arguments about the properties of shapes. As students enter the middle grades, we want to know what they know about these properties.

The following sorting problem, Geodee's Sorting Scheme (developed by the Curriculum Research and Development Group [CRDG n.d., unit 4], can be used to assess students' knowledge by noting what students attend to in explaining Geodee's sorting scheme. It is important to gain as much information as possible about the thinking of students and their ability to classify shapes, to justify their thinking, and to use appropriate geometric terms (see the information in fig. 1.1).

Fig. **1.1.**

Categories of two-dimensional shapes

Shape	Description	Examples
Simple closed curves		
Concave	One or more diagonals of a *concave* figure are outside the figure.	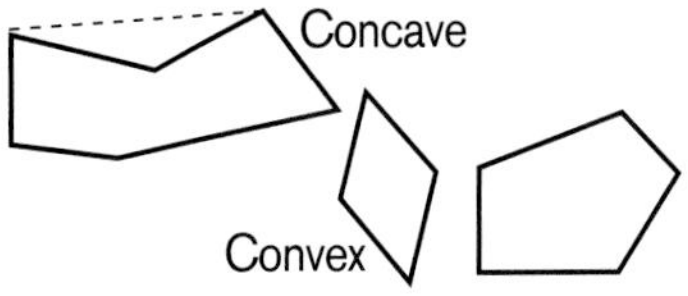
Convex	All diagonals of a *convex* figure are inside the figure.	
Symmetrical	A figure that can be folded so that the two parts match exactly has *line,* or *reflection, symmetry.* A figure that coincides with the original figure after it has been rotated less than 360° has *rotational symmetry.*	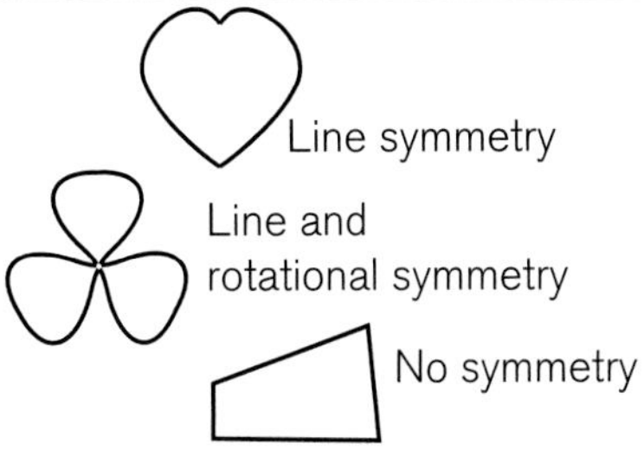
Nonsymmetrical	A nonsymmetrical figure has neither line nor rotational symmetry.	
Polygons		
Regular polygons	*Polygons* are simple closed curves with all straight sides. *Regular polygons* (shaded in green) have all sides and all angles congruent.	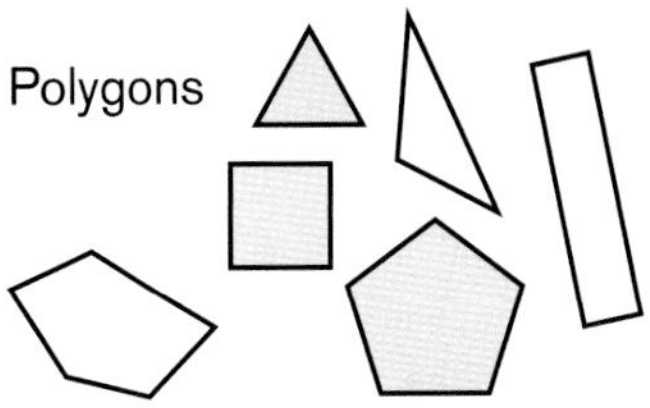
Triangles (classified by properties of sides)		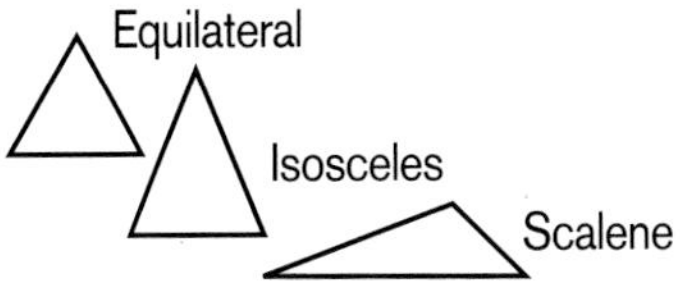
Equilateral	*Equilateral triangles* have all sides congruent.	
Isosceles	*Isosceles triangles* have two sides congruent.	
Scalene	*Scalene triangles* have no sides congruent.	
Triangles (classified by properties of angles)		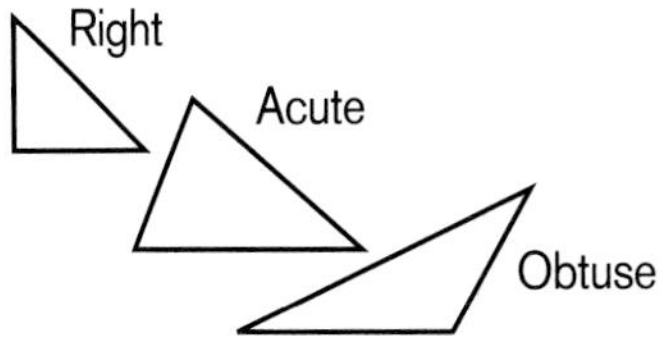
Right	One angle of a *right triangle* is equal to 90°.	
Acute	All angles of an *acute triangle* are less than 90°.	
Obtuse	One angle of an *obtuse triangle* is greater than 90°.	
Quadrilaterals (convex)	A quadrilateral is a polygon with four sides.	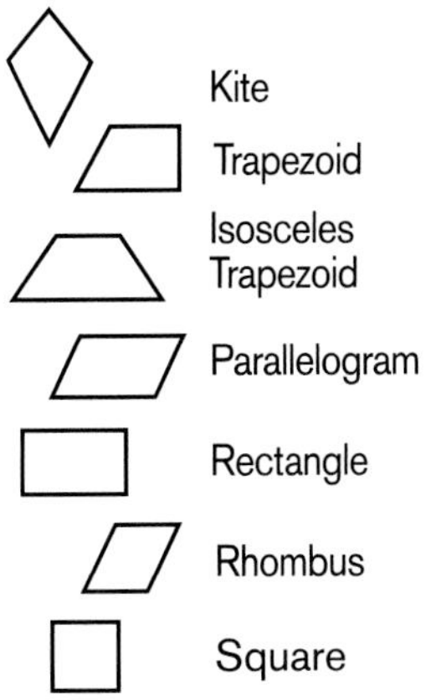
Kite	A *kite* has two pairs of congruent adjacent sides.	
Trapezoid*	A *trapezoid* has exactly one pair of parallel sides.	
Isosceles trapezoid	The opposite nonparallel sides of an *isosceles trapezoid* are congruent.	
Parallelograms	A *parallelogram* is a quadrilateral with two pairs of parallel sides.	
Rectangle	A *rectangle* is a parallelogram with four right angles.	
Rhombus	A *rhombus* is a parallelogram with all sides equal.	
Square	A *square* is a parallelogram with four right angles and all sides equal.	

* Some authors choose to define *trapezoid* as a quadrilateral with at least one pair of parallel sides. That definition is more inclusive and leads to the conclusion that all parallelograms are trapezoids. The Navigations books adopt the classical definition that a trapezoid is a quadrilateral with exactly one pair of parallel sides.

Adapted from Cathcart et al. (2000, pp. 294–95)

Fig. **1.2.**

Categories of three-dimensional shapes

Shape	Description	Examples
Polyhedra	*Polyhedra* are three-dimensional shapes with faces composed of polygons. Polyhedra have faces, edges, and vertices.	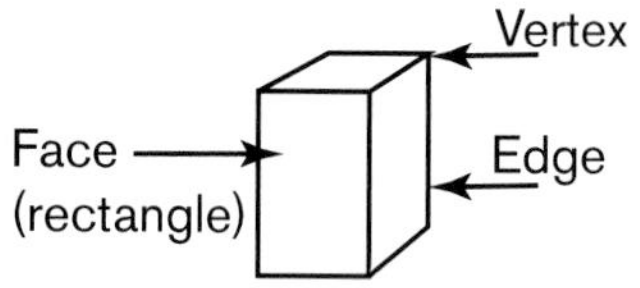 Vertex Face (rectangle) Edge
Regular polyhedra	*Regular polyhedra* have faces consisting of the same kind of congruent regular polygons, and they have the same number of faces meeting at each vertex in the same way. There are five regular polyhedra (also known as Platonic solids).	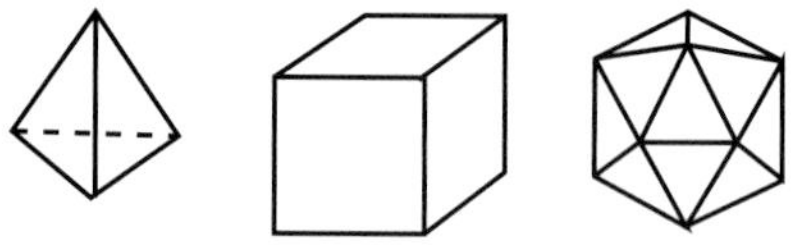
Semiregular polyhedra	The faces of *semiregular polyhedra* consist of more than one kind of regular polygon, and each vertex is surrounded by the same arrangement of polygons. There are thirteen semiregular polyhedra (also known as Archimedean solids).	
Prisms	*Prisms* have two parallel bases that are congruent polygons; the lateral faces are parallelograms formed by segments connecting corresponding vertices of the bases. Prisms are named for the shape of their bases (e.g., trangular prism, square prism, etc.).	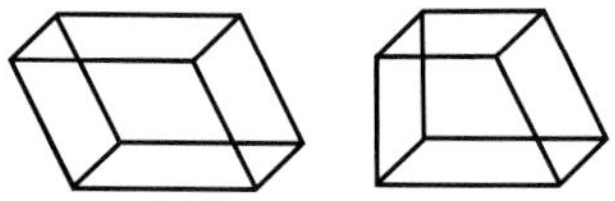
Right prisms	*Right prisms* have lateral faces that are rectangles; the segments connecting corresponding vertices are perpendicular to the bases.	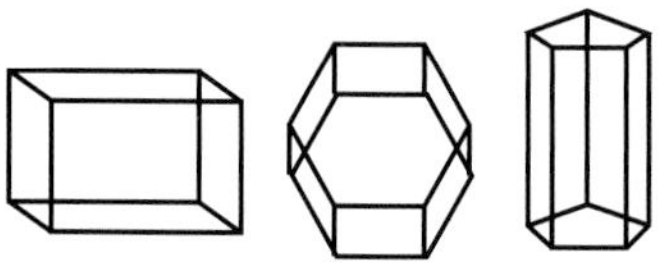
Pyramids	*Pyramids* have bases that are polygons; the lateral faces are triangles that meet at a common vertex. Pyramids are named for the shape of their bases (e.g., triangular pyramid, square pyramid, etc.).	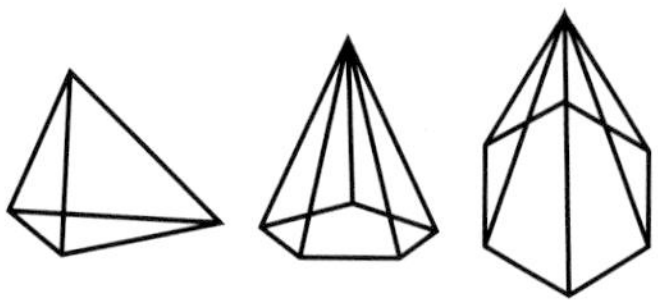
Right pyramids	*Right pyramids* have lateral faces that are isosceles triangles. In a right pyramid, the segment connecting the vertex to the center of the base is perpendicular to the base.	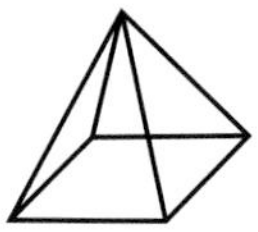
Cylinders	*Cylinders* have curved lateral surfaces joining two parallel bases that are congruent circular regions Segments connecting corresponding points on the bases are parallel.	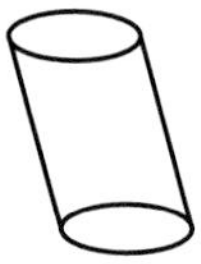
Right cylinders	The segment connecting the centers of the bases (the axis of the cylinder) of a *right cylinder* is perpendicular to the base.	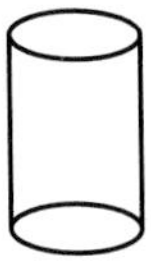
Cones	*Cones* have curved lateral surfaces and bases that are circular regions.	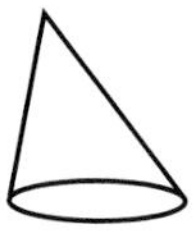
Right cones	The segment connecting the vertex to the center of the base (the altitude of the cone) is perpendicular to the base.	

Adapted from Cathcart et al. (2000, pp. 286–88)

Geodee's Sorting Scheme

Goal

To assess students'—

- ability to recognize characteristics and properties of two-dimensional shapes;
- ability to recognize characteristics and properties of three-dimensional shapes;
- ability to justify their geometric thinking.

Materials and Equipment

- A copy of the blackline master "Geodee's Sorting Scheme" for each student (or pair of students)
- A set of three-dimensional figures, including regular and semiregular polyhedra, cylinders, prisms, cones, and pyramids

p. 86

Activity

In this activity, the students first work alone or in pairs to explain Geodee's method for separating shapes into two categories and to classify one remaining shape. Distribute the activity sheets, and allow ample time for the students to examine the diagram of Geodee's two categories of shapes. You might encourage them to think about how the two categories of shapes are alike and how they are different.

Discussion

After the students have had time to formulate arguments for the questions on the blackline master, allow them to share their responses in pairs or small groups and justify their decisions. Extend the conversation to a whole-class discussion by selecting volunteers to share their thinking. Encourage the other students to comment in a nonthreatening manner on the strengths and weaknesses of the volunteers' arguments.

The Shape Sorter applet on the CD-ROM could be used in conjunction with this activity.

Geodee's problem extends the students' analysis to require them to create and justify their definitions for the categories. Shape O could be placed in either category, depending on how the classification scheme is defined. If category 2 contains regular polygons, the shape goes in category 1. If category 2 consists of polygons with congruent sides, shape O could be placed in it. This task sequence requires the students to focus their attention on certain properties and to be more precise in their analysis and descriptions.

Extensions

The following "clue" task, in which students use clues to determine a mystery shape, provides further explorations with the characteristics and properties of two-dimensional shapes. This task gives students an opportunity to apply their understanding. Clues for only two shapes are provided, but teachers are encouraged to construct other sets of clues according to the shapes the class is or has been studying.

Read the clues below; they describe two different mystery shapes. Name the shapes described and draw them. Explain how the clues fit the shape you have drawn.

Clues for Shape A	Clues for Shape B
• Four sides • Two pairs of sides equal in length • No lines of symmetry	• Exactly three lines of symmetry • No right angles or obtuse angles • All sides of equal length

Shape A is a parallelogram that is not a rectangle or a rhombus.

Shape B is an equilateral triangle.

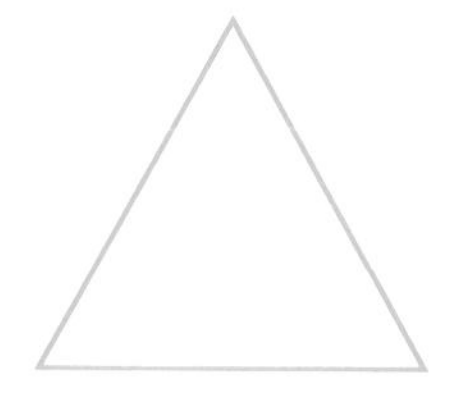

The applet Spinning and Slicing Polyhedra can help students enhance their understanding of three-dimensional shapes.

In other tasks, the students can be challenged to draw or construct as many different shapes as possible that exhibit certain given properties. For example:

- Make four-sided shapes in which both pairs of opposite sides and opposite angles are equal.
- Make four-sided shapes that have more than one line of symmetry.
- Make four-sided shapes with one pair of opposite sides parallel.

Other challenges include these:

- Can you make a triangle with two right angles?
- Can you make a parallelogram with only two equal sides?

Similar activities can be conducted using three-dimensional shapes to determine students' understanding of this dimensionality (see the information in fig. 1.2). It is helpful to have available a set of three-dimensional figures, including regular and semiregular polyhedra, cylinders, prisms, cones, and pyramids. Then students can be asked to do tasks like the following:

- Choose two shapes, and tell how they are alike and how they are different.
- Sort a collection of shapes into two different sets, and explain why the shapes in each set belong together.
- Examine a shape and write several statements about its characteristics.

When students do each activity, what do they notice? Do they pay attention to the presence or absence of edges and vertices? Do they focus on faces and surfaces? Do they distinguish polyhedra, cylinders, and cones? It is likely that students entering the middle grades will not have had as many experiences with three-dimensional shapes as with two-dimensional shapes.

Selected Instructional Activities

Good tasks such as the ones you will see throughout this book provide means for both instruction and assessment—ways for students to develop their understanding and ways for teachers to learn more about how their students think and what they understand. Explorations that focus on the characteristics and properties of shapes can be enhanced with dynamic geometry software that helps students visualize and understand relationships, thus providing a necessary foundation for the development of geometric reasoning. Although such explorations could be done by hand, dynamic geometry software allows students to alter shapes and observe the resulting effects, as in the next activity, Exploring Triangles (adapted from PBS TeacherSource [PBS 2001]). Dynamic geometry programs can also be used to launch rich explorations of relationships among different shapes.

Exploring Triangles

Goals

- Explore the measures of angles of triangles
- Explore the lengths of sides of triangles
- Explore the relationships between the measures of angles and the lengths of sides of triangles

Materials and Equipment

- Dynamic geometry software and enough computers to accommodate no more than two or three students on each one. If such software is not available, the activity can be modified by having students draw several triangles and vary the location of point *C*, as specified in each part of the activity.
- A copy of the blackline master "Exploring Triangles" for each student or group of students

p. 87

Activity

The students use dynamic geometry software to construct a triangle *ABC*, then measure all the angles and the lengths of the sides. Figure 1.3 gives an example of such a triangle. The students then observe what happens to the measures of the angles and the lengths of the sides as point *C* is moved.

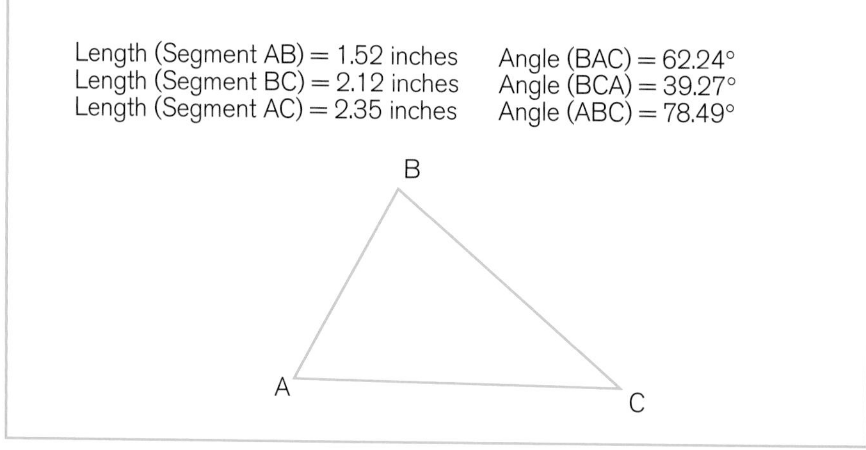

Fig. **1.3.**
A triangle *ABC*

Discussion

The goal of this activity is to help students discover the relationship between the measures of the angles and the lengths of the sides of triangles. Dynamic geometry software allows great flexibility for students to change various properties of figures and observe the effects of those changes. This flexibility is a powerful tool for exploring the relationships among properties of shapes. It is important for students to formulate ideas related to the exercises. You might pose questions and have the students explore them in small groups or as a whole class. For example, in question 2 the students should reflect on why angle *C* becomes larger. One possible response is that as $\overline{AB}$ becomes the longest side of the triangle, angle C becomes the angle with the largest measure. The third question helps students visualize that when angle C become coincident with side AB, it has a measure of 180 degrees. The students might discuss that when C lies on $\overline{AB}$, a straight angle is formed and $\overline{AC} + \overline{CB} = \overline{AB}$. Question 4 reinforces important ideas about types of triangles. You might ask the students to discuss what changes would produce an isosceles triangle (two congruent angles with congruent sides opposite those angles) or an obtuse triangle (having one angle between 90° and 180° with the side opposite that angle having the greatest measure). Question 5 should prompt a summary of the important ideas that students should have discovered. You might ask the students to demonstrate their ideas with more than one figure, or you could have groups or individuals share examples. The students could thus see that although figures may be different, the relationships being emphasized remain the same.

You might also ask the students to compare the sum of the lengths of two sides of a triangle to the length of the third side to acquaint them with the triangle-inequality theorem: The sum of the lengths of any two sides of a triangle is greater than the length of the third side.

Understanding Relationships among Similar Objects

In addition to their explorations of properties of shapes, middle-grades students also need experiences in working with congruent and similar shapes. From their work in grades 3–5, students should understand that congruent shapes can be "matched" by placing one atop the other. Similar shapes have congruent angles but not necessarily congruent sides, so when one similar figure is placed atop another, the two may not "match" exactly in size.

Two basic questions students must answer when studying congruence and similarity are (1) What is meant by the same shape? and (2) What characteristics of two shapes must be alike before two figures can be considered the same shape? You may wish to reproduce the blackline master "Congruent and Similar Shapes" (adapted from O'Daffer and Clemens [1992, pp. 250–51]) as an overhead transparency to use in introducing the concepts of similarity and congruence. In the elementary grades, students study congruence by exploring and identifying the ways in which shapes are alike and by showing concretely that one shape can be made to coincide with another. Often, intuition helps students identify objects that are congruent. In the middle grades, however, students are expected to go beyond identification to formulate and critique arguments about congruence and similarity.

p. 89

At this level, the study of similarity involves a more-complex analysis of shapes and requires the ability to recognize and understand multiplicative relationships. In the middle grades, one method students use to analyze two shapes to determine similarity involves measuring sides and angles. Students find that in similar shapes, corresponding angles have the same measure and the lengths of corresponding sides are proportional. Students who are not skillful in measurement techniques may get confounding results and thereby miss finding important relationships among shapes. Another method that gives students experience with similar shapes uses dilations. Dilations are nonrigid motions that result in enlargements and reductions of shapes. The

"In the middle grades, [students] should extend their understanding of similarity to be more precise, noting, for instance, that similar shapes 'match exactly when magnified or shrunk' or that their corresponding angles are congruent and their corresponding sides are related by a scale factor."
(NCTM 2000, p. 234)

following series of problems is designed to move students from discussing dilations intuitively to employing a technique that more precisely dilates shapes in order to study the relationships among similar shapes.

p. 90

The elephant problem (developed by CRDG, University of Hawaii) is presented in the blackline master "Comparing Elephants." It allows students to use their intuition to describe the differences and similarities between figures. Reproduce the blackline master on a transparency for class discussion, or make paper copies for students to discuss in pairs or small groups. The goal of the exploration is to help students better understand the concepts of similarity and congruence before they begin more-formal explorations with dilations.

The students may notice that some of the elephants seem to be the "same" as elephant A but a different size (elephants E and I); some are the exact size (elephants C and G). Some students may use the word *proportional* to describe these elephants. If so, ask them what they mean by the term. The students should understand that figures are congruent only if they are *exactly* the same size and shape, although one shape may have been translated, rotated, or reflected relative to another shape. Figures that have the same shape but differ in size are similar.

The next problem (also developed by CRDG, University of Hawaii), which involves drawing an enlargement, takes the students' intuitive notion of similarity one step further.

Exploring Similar Figures

Goals

- Enlarge figures (use dilations)
- Understand similarity and proportionality

Materials and Equipment

- A copy of the blackline master "Exploring Similar Figures" for each student or pair of students
- Rulers
- Protractors
- Calculators

p. 91

Activity

Distribute a copy of the activity sheet to each student or pair of students. In this task, the students are required to complete the enlargement of a heptagon.

Discussion

You can begin a discussion of this problem by asking the students, "What were you trying to accomplish as you made the enlargement?" As the students talk about trying to keep the angles congruent and the sides proportional, the question serves to highlight the characteristics of a proportionally enlarged figure. The ratio of the lengths of the corresponding sides of the figures is called a *scale factor*. Some students may even mention that the corresponding sides of the original image and the new image are parallel, a unique feature of dilation images. If the students do not mention specific characteristics, prompt such analysis with questions such as What changed in the enlarged figure? What stayed the same?

Extension

Dynamic geometry software can help students explore the similarity of figures. Students can construct similar figures, obtain the measures of several of the sides and angles, and then find the scale factor for the figures. It is important for students to understand that similar figures have corresponding sides that are proportional and corresponding angles that are congruent (see fig. 1.4). Note that the scale factors are reciprocals (i.e., the scale factor for the enlargement is the reciprocal of the scale factor for the reduction). Students should also realize that the scale factor depends on which figure is the original, that is, on what is defined as 100 percent.

In the following activity, Dilating Figures, students can continue their study of similarity by learning to do dilations.

Fig. **1.4.**

Parallelogram *STUV* is similar to parallelogram *CDEF*.

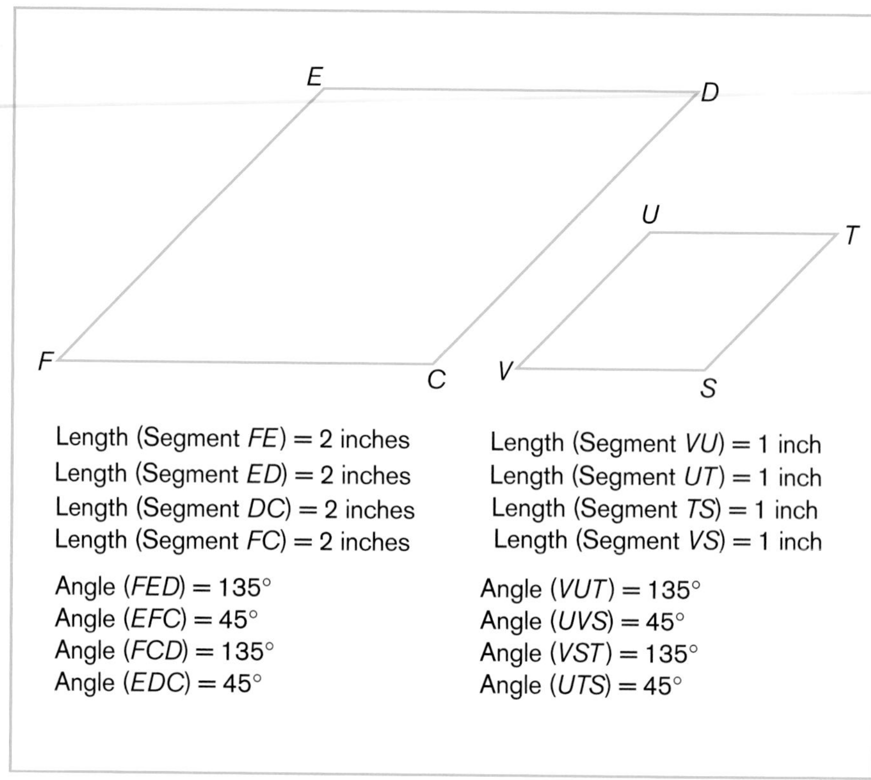

Dilating Figures

Goals

- Use dilations to draw similar figures
- Recognize similar figures and corresponding parts

Materials and Equipment

- A copy of the blackline master "Dilating Figures" for each student or pair of students
- Rulers or protractors

p. 92

Activity

Distribute a copy of the activity sheet to each student. In the problem, a pentagon and Sheri's incomplete dilation of the pentagon are displayed, and the students are asked to determine the method Sheri is using to perform the dilation.

Discussion

Use the following questions to help structure a discussion through which students can reflect on the activity:

- How is Sheri using point T inside $ABCDE$? (Point T is the center of dilation. From this point, rays are extended through the vertices of $ABCDE$ to obtain the points for the vertices of the dilated pentagon.)
- How is Sheri drawing segments to complete the dilation? (Sheri is drawing segments by completing the dilation lines from T through all the vertices.)
- How is Sheri using these segments? (The pentagon has been enlarged to 200 percent of its original size. In this example, the distance from T to A' is twice the distance from T to A. Another way to view the change is as a scale factor of 2. The rays are drawn from point T [the center of dilation] through the vertices of the original image.)
- How did Sheri determine where to make the hash marks? (The hash marks are made at the points whose distances from T are 200% of [twice] the lengths of the segments from the center to the corresponding vertices of the original pentagon. Note the question above.)
- How did Sheri use the hash marks? (They become the vertices that are connected to form the dilated pentagon $A'B'C'D'E'$.)
- What were some things you noticed about the two figures? (Many answers are possible.)

In a follow-up discussion, focus on—

- the relationship of the corresponding sides and of the angles (you may want the students to measure them to see that the corresponding angles in the original and the dilated images are congruent and that the corresponding sides are proportional);

- the relationship between the perimeters (ask the students if it corresponds to the scale factor);
- the orientation of the figures (mention that segments and their images under a dilation are parallel, e.g., side AB is parallel to side $A'B'$);
- the preservation of collinearity—that is, if P is a point on segment AE in the original figure, then P' is on segment $A'E'$ in the dilation;
- what it means to dilate a figure 200 percent. The dilated figure is 200 percent of the size of the original figure. The students might also say the scale factor is 2. If the scale factor is greater than 1, the dilation is an *enlargement;* if the scale factor is less than 1, it is a *reduction.* If the scale factor equals 1, the preimage and the image are the same; that is, the dilation is an isometry. This concept is important for students to understand.

To emphasize these relationships, you might have the students use the same diagram to create a dilation that is 50 percent of the original pentagon $ABCDE$.

Creating and Critiquing Inductive and Deductive Arguments

Tasks that develop geometric reasoning are cognitively demanding and require students to explore as well as understand the nature of relationships (Stein, Smith, Henningsen, and Silver 2000). The previous problems illustrate the types of experiences that students must have in order to develop an understanding of the relationships among, and properties of, shapes. As students conjecture about, test, and discuss hypotheses, they are developing thinking skills that are necessary to structure more-formal arguments. Activities designed to develop geometric thinking should—

- stress mathematical reasoning and argumentation;
- provide opportunities to formulate and test conjectures about geometric relationships;
- require both informal and formal proofs as students construct convincing mathematical arguments.

The next activity illustrates how Venn diagrams can help students clarify their reasoning about shapes.

Using Venn Diagrams to Reason about Shapes

Goals

- Explore different ways to classify triangles
- Extend students' thinking about relationships among various types of triangles

Materials and Equipment

- A copy of the blackline master "Using Venn Diagrams to Reason about Shapes" for each student or pair of students

p. 93

Activity

Distribute an activity sheet to each student or, if you prefer to conduct the activity in groups, to pairs of students. On a Venn diagram consisting of two intersecting circles, the students label sets A and B with types of triangles and explain why they did so. They then draw examples of triangles that would be found in each of the circles and in the intersection of the circles and explore the relationships among the three sets of triangles.

Students can also do this activity using the Shape Sorter applet on the CD-ROM.

Discussion

Venn diagrams provide a visual organizer to help students consider relationships. The goal of this task is to help students formulate arguments to justify their placement of various triangles in the Venn diagram. You may need to explain Venn diagrams to the students. Many students are particularly concerned about region 2, where any triangle listed must have the characteristics of the triangles in both area 1 and area 3. For example, a student might label set A "equilateral triangles" and set B "isosceles triangles." The students' drawings should show all equilateral triangles placed in region 2 (the intersection). Their justifications should point out that all equilateral triangles can also be classified as isosceles triangles and that therefore no examples of equilateral triangles should be placed in region 1 but that some isosceles triangles (those with only two congruent sides) are not equilateral triangles and that those triangles should appear in region 3. Encouraging the students to illustrate their reasoning with diagrams can help them articulate their arguments.

You might also consider adding a fourth area, representing everything outside sets A and B. Label the area outside the circles "area 4." This area would give students the option of identifying triangles that do not fit the characteristics for either set A or set B. In the previous example, a student might draw a scalene triangle in area 4, since scalene triangles (triangles with no sides congruent) cannot be either equilateral or isosceles.

A variation of this Venn-diagram activity is to use types of quadrilaterals, such as rectangles, squares, rhombuses, trapezoids, or kites. This

Questioning students' responses and giving students opportunities to revise their answers can aid in developing their ability to communicate and reason about geometric properties.

variation helps students focus on both the angles and the sides of the shapes. It is important always to have students justify their selections.

Consider how some eighth-grade students responded to this problem and how their teacher addressed responses in a way that encouraged additional thinking about the relationships. Jessica listed isosceles triangles in set A and right triangles in set B. Initially, her explanation did not provide any information about the types of triangles she would display in area 2. The teacher prompted her to think about the characteristics of right triangles and of isosceles triangles. Jessica's follow-up discussed that "triangles in area 2 would be both isosceles and right triangles. They would have a 90° angle and two sides that are the same size. This area shows a special kind of triangle." Guillermo placed isosceles triangles in set A and right triangles in set B. He said that all right triangles are in region 2, affording the teacher an opportunity to challenge him to consider the relationship. The teacher asked him what the diagram says about all right triangles. Since region 2 is within region 1 (that is, set A), Guillermo's response implied that all right triangles are isosceles triangles. The Venn diagram served as a tool that helped both students focus their thinking about relationships and develop logical-reasoning skills.

What does Guillermo's statement imply?

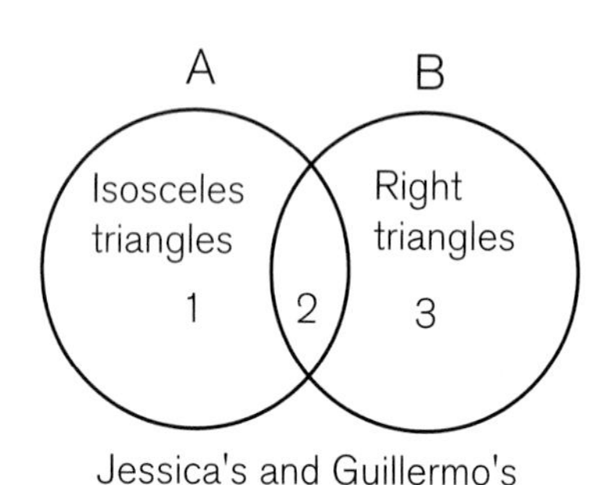

Jessica's and Guillermo's problem

Tools such as Venn diagrams can be helpful in giving students a structure for reasoning about the properties of figures and the relationships among them. They support the development of inductive reasoning, which involves making predictions on the basis of data and observed patterns. Such conclusions are likely to be true. They are not proofs but rather conjectures based on observations.

The next activity, Midsegments of Triangles, engages students in collecting data and making conjectures about geometric relationships on the basis of the data collected. Conjecturing and creating arguments based on data are indicative of inductive reasoning.

Midsegments of Triangles

Goals

- Locate the midpoints of the sides of triangles
- Construct midsegments of triangles
- Make and test conjectures

Materials and Equipment

- A copy of the blackline master "Midsegments of Triangles" for each student
- Paper for drawing
- Rulers for measuring
- Calculators
- Dynamic geometry software (optional)

p. 94

Activity

The students draw five triangles by hand or using dynamic geometry software. If the triangles are drawn by hand, the students may experience difficulty in obtaining precise measurements. You may want to forewarn the students of this potential difficulty.

Next, the students locate the midpoints of two sides of each of the five triangles and then in each triangle draw the line segment connecting the midpoints, which forms the midsegment of the triangle. The students then measure and record in a table the lengths of the midsegments and the lengths of all the sides of the five triangles, look for patterns in their data, and describe any patterns that emerge.

Dynamic geometry software allows students to investigate the relationship of the midsegment to the sides without having to use rulers to determine the measures. Such software can provide measures of the sides, thus allowing the students to focus on identifying and describing relationships. See figure 1.5.

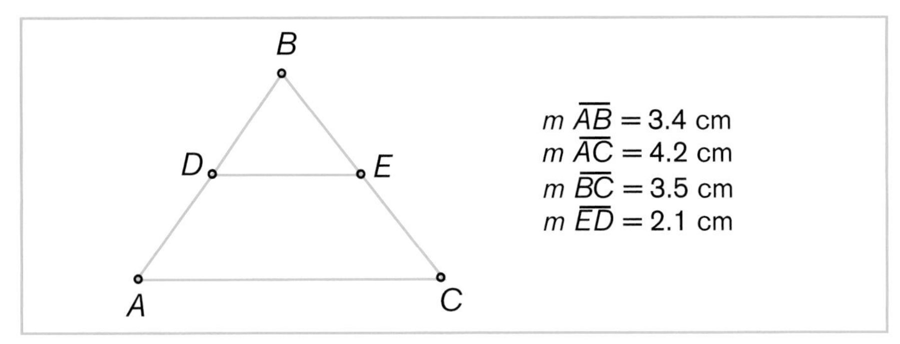

Fig. **1.5.**

A triangle and midsegment drawn with dynamic geometry software

Discussion

The students' measurements will vary; however, their data should lead them to the conjecture that the length of a midsegment in any triangle is always half the length of the third side of the triangle (the side parallel to the midsegment). You might also ask the students to look at the relationships of the angles of the original triangle to the angles formed by the midsegment and the sides of the triangle. If the students have already studied parallel lines, then this task would be a good review of the concept. If the students have not studied parallel lines,

Figure 1.6 summarizes the relationships of the measures of the angles formed when two parallel lines are cut by a transversal. Note that the corresponding angles (angles 1 and 5, 2 and 6, 3 and 7, and 4 and 8), the alternate interior angles (angles 3 and 6 and angles 4 and 5), and the alternate exterior angles (angles 1 and 8 and angles 2 and 7) are congruent.

Fig. **1.6.**

The angles formed when two parallel lines are cut by a transversal

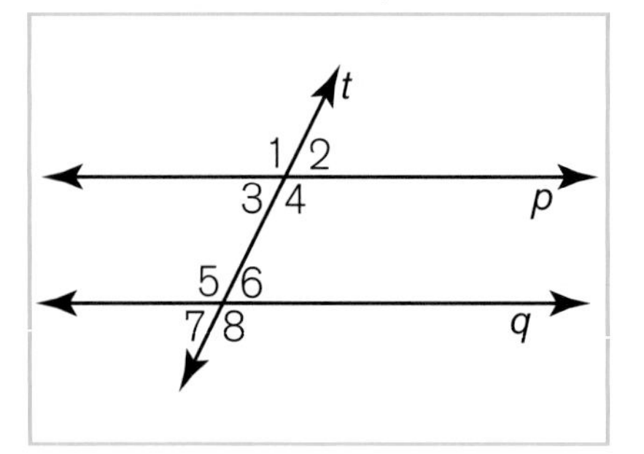

considering the relationship of these angles emphasizes an important geometric property: If parallel lines are cut by a transversal (in this instance, the sides of the triangle), then the corresponding angles are congruent. Dynamic geometry software allows students to select three points that define an angle and obtain the measure of that angle.

This activity illustrates *inductive reasoning* (making conjectures on the basis of observations and data). *Deductive reasoning* begins with a statement that is accepted as true and demonstrates that other statements follow from the original assumption. In applying deductive reasoning, students would *prove* that a midsegment of a triangle is *always* half the measure of the third side by drawing on definitions, theorems, and postulates. Student would need to develop a logical line of reasoning to establish that the conjecture is *always* true. This degree of rigor is not expected of middle-grades students; engaging students in making conjectures, however, will draw them into making convincing arguments and finding counterexamples to demonstrate the flaws in conclusions. Such reasoning supports them in making connections between the data and the conjecture when they examine geometric relationships and helps them make a successful transition from inductive reasoning to deductive reasoning, on which formal proofs depend.

Another way to build reasoning skills and to extend students' knowledge about the characteristics and properties of shapes is to have students explore statements in forms such as "If ___, then ___" and "All ___ are ___." The blanks can be filled in with the names of various geometric shapes, properties, or relationships. The following are examples of such statements (adapted from Van de Walle [1998, p. 379]), of which some are true and others are false:

- If it is a square, then it is a rectangle.
- If it is a rectangle, then it is a square.
- If it is a square, then it is a rhombus.
- If it is a right triangle, then it is an equilateral triangle.
- All parallelograms have congruent diagonals.
- All quadrilaterals with congruent diagonals are parallelograms.
- If two rectangles have the same area, then they are congruent.
- If two squares have the same perimeter, then they are congruent.
- If a prism has a plane of symmetry, then it is a right prism.
- If a hexagon can be divided into six equilateral triangles, then it is a regular hexagon.

Students, working alone or in groups, should be encouraged to use drawings or models to reason about the validity of such statements. When working in groups, students should develop convincing arguments that demonstrate whether statements are true or false. The use of counterexamples gives students opportunities to use logic to support their positions. Students might also be encouraged to compose similar statements and develop lines of reasoning to support them.

Students' reasons may include some of the following statements:

- A rectangle is a parallelogram with four right angles, a square is a parallelogram with four right angles and four congruent sides, and a rhombus is a parallelogram with four congruent sides.

- An equilateral triangle has three congruent sides and three congruent angles (so each must be 60 degrees); a right triangle has one angle that is 90 degrees.
- If the diagonals are congruent, a parallelogram is a rectangle.
- Rectangles can have equal areas but different dimensions; for instance, a six-inch-by-two-inch and a three-inch-by-four-inch rectangle have the same area but are not congruent.
- Squares have four congruent sides, so the perimeter is $4s$, where s is the length of one side; so if the perimeters are equal, then the squares are congruent.
- A prism is a right prism if all its lateral faces (not the bases) are rectangles; otherwise, it is an oblique prism. Its bases are congruent polygons lying in parallel planes.
- If a hexagon can be divided into six equilateral triangles (assuming nothing is left over), then it is a regular hexagon.

One of the most important relationships middle-grades students should become familiar with is the Pythagorean theorem. Students should be familiar not just with the algebraic model of the relationship. They should also develop a geometric understanding of it and be able to apply the theorem to relevant problems. The following activity, Reasoning about the Pythagorean Theorem, uses an area model to develop an understanding of the Pythagorean theorem.

Reasoning about the Pythagorean Theorem

Goals

- Become familiar with the Pythagorean theorem
- Use an area model to discover the Pythagorean theorem

Materials and Equipment

- Two or more sheets of grid paper (available on the CD-ROM that accompanies this book) and one ruler for each student
- A copy of the blackline master "Reasoning about the Pythagorean Relationship" for each student

p. 95

Activity

The students draw two right triangles, one with sides measuring three, four, and five units, and the other with sides measuring five, twelve, and thirteen units. On each triangle, they draw squares, using the sides of the triangle as sides of the squares. For each triangle, they find the area of the squares and enter their results in a table, along with the measures of the sides of each triangle.

Discussion

This exploration should be continued with two or three more triangles. You might model another example for the students so you can address precision in measurement (many teachers prefer to use metric units). In cases where nonintegral measures occur, rounding may be necessary to make the Pythagorean relationship clear. For example, a triangle with legs measuring 2 centimeters and 3 centimeters has a hypotenuse of approximately 3.6 centimeters. The students may continue to draw the triangles on graph paper, but they will need to use a ruler to measure the sides and compute the areas of the squares. Some students may focus more on completing measurements than on the relationship among the squares of the sides. The students may need to be prompted specifically to look for patterns in the data that indicate relationships among the parts of the figures.

A seventh-grade teacher reported using Cuisenaire rods to demonstrate the Pythagorean relationship, as shown in figure 1.7.

Teachers recount that although many students may be able to manipulate the formula $c^2 = a^2 + b^2$ and apply the Pythagorean theorem

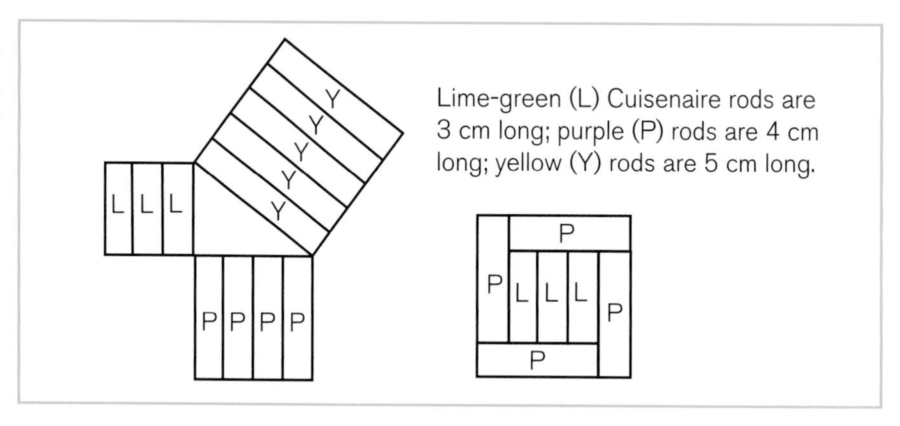

Fig. **1.7.**
The Pythagorean relationship demonstrated with Cuisenaire rods

appropriately to situations, they do not have a deep understanding of the relationship. Therefore, multiple approaches should be used to allow students to explore this important relationship in depth. The students will encounter this theorem, which has a rich history, again and again. Those who are able to develop a sound rationale for why the formula works are demonstrating reasoning that will continue to be developed through more-formal studies of geometry. Students should be encouraged to engage in oral and written communication about their reasoning. The following questions can stimulate some fruitful thinking:

- What conclusion can you make on the basis of your work with right triangles?
- How do the various tasks you have undertaken support the Pythagorean theorem?

What does this theorem say about the possibility of a right triangle with sides measuring 3, 5, and 7 units? Support your conclusion with an illustration and an argument.

Exploring Relationships with Polyhedra

The next task continues students' investigations of the properties and characteristics of shapes by focusing on three-dimensional shapes. It requires students to form conjectures about relationships on the basis of data and connects algebra with geometry.

Have the students use nets to construct several polyhedra, including a cube, a pentagonal pyramid, and a square pyramid. Ask the students to count the vertices, faces, and edges of each polyhedron. A table such as the one in figure 1.8 will help the students keep track of their observations.

The CD-ROM that accompanies this book contains nets of polyhedra, including the five regular polyhedra, which can be used with this and other geometric explorations.

Name of Polyhedron	Number of Faces (F)	Number of Vertices (V)	Number of Edges (E)
Tetrahedron	4	4	6
Hexahedron (cube)	6	8	12
Octahedron	8	6	12
Dodecahedron	12	20	30
Icosahedron	20	12	30
Pentagonal prism	7	10	15
Triangular prism	5	6	9
Pentagonal pyramid	6	6	10
Square pyramid	5	5	8

Fig. **1.8.**

An example of a chart in which students have recorded their observations about the properties of polyhedra

Ask the students to study their data carefully in order to detect any patterns in the numbers of faces, vertices, and edges. They should begin to uncover Euler's formula, which can be expressed in varying forms. The formula Vertices (V) + Faces (F) = Edges (E) + 2 is perhaps the most common. Sometimes, the relationship is written as $V + F - E = 2$. The formula holds for both regular and irregular convex polyhedra. Allow time for the students to explore the relationship among the faces, vertices, and edges and express this relationship in words and in more-formal expressions such as those above. Your students might enjoy using

Allowing students to develop conceptual meanings for definitions of various polyhedra assists in developing a broader and more flexible understanding of important geometric terms. See Jane Keiser (2000), "The Role of Definition," included on the CD-ROM.

Counting Parts of Solids on NCTM's Illuminations Web site to explore these relationships (address: www.illuminations.nctm.org; click on i-Math Investigations, and scroll down to the section for grades 3–5).

You could also demonstrate this relationship using a potato or a piece of Styrofoam. Randomly cut flat slices of potato until a flat-faced polyhedron emerges. (You could also cut potatoes ahead of time at home and then have the students use them in small groups.) Count the faces, edges, and vertices. Have the students keep a record of the information they collect. Point out that the totals for an irregular polyhedron fit Euler's formula. The students could also build various irregular polyhedra themselves and verify that the formula holds. Another extension of the task is to have the students truncate (cut off) a corner of one of their polyhedra and discover that the formula still holds.

Conclusion

The activities, tasks, and examples in this chapter were selected to illustrate ways to provide students with important experiences that extend their ability to analyze the characteristics and properties of shapes and to use geometric knowledge to make conjectures and reason effectively. The activities address important goals from *Principles and Standards for School Mathematics* (NCTM 2000): understanding relationships among two- and three-dimensional shapes, understanding relationships among similar figures, and developing inductive and deductive arguments. The instructional goal of these activities and tasks should be to encourage students to use sound reasoning based on the characteristics and properties of shapes as they make conjectures and construct arguments.

GRADES 6–8

NAVIGATING *through* GEOMETRY

Chapter 2

Coordinate Geometry and Other Representational Systems

Use coordinate geometry to represent and examine the properties of geometric shapes

Use coordinate geometry to examine special geometric shapes, such as regular polygons or those with pairs of parallel or perpendicular sides

Important Mathematical Ideas

According to *Principles and Standards for School Mathematics* (NCTM 2000), students in grades 6–8 should use coordinate geometry to represent and examine the properties of geometric shapes, including special figures such as regular polygons or those with pairs of parallel or perpendicular sides. Representational systems (e.g., the coordinate plane) help situate discussions of movement and change in both geometry and algebra. Students may also use such systems to describe line and rotational symmetry as well as flips, slides, and turns. Explorations with the coordinate plane, therefore, are significant in that they provide a context for students to further their understanding of transformations, similarity, and congruence.

What Might Students Already Know about These Ideas?

In the elementary grades, students explore movement, location, direction, and distance. They do so in part by using coordinate grids and numeric coordinates to label and locate points. Their explorations may also include notions of the symmetry, congruence, and similarity of shapes drawn on a coordinate grid, and they may have expanded their investigations to include work in all four quadrants of the Cartesian system. Most students enter the middle grades with some common language and geometric vocabulary to describe location and movement on a

The CD-ROM contains a blackline master for grid paper that can be used for activities involving coordinate systems.

coordinate grid, and they are likely to be able to describe paths and distances between pairs of points with the same vertical and horizontal coordinates.

As a preassessment, students can be asked to draw the coordinate plane on a piece of grid paper and then plot and label a series of points that, when connected, make their initials. This brief task will allow the teacher to evaluate the students' existing knowledge. You may wish to think specifically about the following questions:

- Do the students initially include only positive coordinates on the grid?
- Do they label the axes appropriately?
- Can they locate the x- and y-coordinates?
- Do they plot points in more than one quadrant?
- Can they name points correctly?

This task can yield important information about students' rudimentary understanding of, and experiences with, coordinate geometry.

"Geometric and algebraic representations of problems can be linked using coordinate geometry." (NCTM 2000, p. 235)

Selected Instructional Activities

Building on the experiences students have had in the elementary grades, the activities in this chapter provide rich opportunities to extend students' understanding of coordinate geometry.

The assessment task described above can be extended in numerous ways with the goal of giving students engaging experiences in using the coordinate plane. For example:

- Start by naming the coordinates of the vertices of a polygon. Triangles, rectangles, and squares are accessible starting places. Before the students plot the points on a coordinate grid, ask them to infer (and subsequently write) as much as possible about the polygon in question simply by examining the points presented to them. For example, consider the set of points (–1, 5), (6, 5), (–1, 9), (6, 9), which form the vertices of a rectangle. The students might notice that two of the points have the same x-coordinate and therefore lie on the same vertical line. Or perhaps they will notice that the points reside in different quadrants.

- Give the students the coordinate points for three vertices of a triangle. Encourage them to determine as much as possible about the triangles *prior* to plotting the coordinates. If you give them (2, 1), (4, 1), and (2, 3), for example, do they identify the points as vertices of an isosceles triangle? Of a right triangle? Can they give the lengths of any of the sides (such as two units for the length of the legs)? Do they know that the area is two square units? It is important to require the students to explain how they got their answers.

- Give the students the coordinates of three points that *do not* form a triangle (i.e., that lie on the same line). You might select points such as (–1, –1), (2, 2), and (6, 6). Challenge the students to explain the difference between these points and those that formed the vertices of the triangle explored earlier.

Students may enjoy using the Coordinate Geometry Tool applet on the CD-ROM to accomplish many coordinate geometry tasks.

- Describe a polygon (start with a rectangle) by giving the coordinates of all the vertices except one. Give the students time to determine the remaining vertex, and then check their thinking by having them represent the figure on grid paper. If you give them points (–1, 2), (–1, 6), and (1, 6) for a rectangle, for example, can they determine that (1, 2) is the missing point? Can they justify their thinking?

- After the students have represented and explored a particular polygon in the coordinate system, encourage them to determine coordinates that will result in a shape that is *congruent* to the first polygon. (You may want to provide them with one point to orient the new figure.) This activity becomes more challenging when the students create congruent figures in quadrants other than the quadrant in which the original figure lies or when they represent congruent figures that lie in two or more quadrants. These activities provide evidence about whether students understand congruence and whether they can subsequently determine appropriate distances between the vertices of a polygon.

- Extend the previous activities involving congruence by encouraging the students to think likewise about *similarity*. For example, the students could be given the coordinates for the vertices of a triangle and asked to determine coordinates that would produce a triangle that is similar to the original triangle.

- If repeated experiences with figures whose bases are parallel to the x-axis have reduced the challenge for students, change the orientation of the figures. For example, giving (1, 3), (3, 4), (1, 1), and (–1, 0) as coordinates of the vertices of a polygon may stimulate students to engage in more-complex thinking about the properties of a parallelogram.

Teachers should vary the orientation of figures in coordinate space so that students are not bound by orientation in identifying and analyzing polygons.

Throughout these activities, the students should share their thinking and strategies, particularly when several different solutions are appropriate.

From the previous tasks, it can be seen that the coordinate plane is a meaningful context in which students can examine properties of various shapes. The coordinate plane can be used to highlight the primary properties that distinguish one shape from another. In addition, using rudimentary algebraic tools and the Pythagorean relationship, students can use the coordinate plane to compute the distances between the vertices of a polygon (the length of the sides) or to compute the slope of line segments to determine parallelism. They can also explore angle measures informally by closely examining the relationship among three points on the coordinate plane. Such experiences promote concepts that are important for developing a broad sense of geometry in students and illustrate the connections between algebra and geometry.

On the CD-ROM, see "Using Communication to Develop Students' Mathematical Literacy" (Pugalee 2001), which describes the power of communication in extending students' thinking about mathematics.

Several tools are available to facilitate students' explorations on the coordinate plane. Some handheld calculators allow the user to turn on a grid system. Check the manufacturer's Web site to see if lessons or activities are available online. Dynamic geometry programs allow greater flexibility, with features that display coordinates of points and measures of attributes such as distance, length, radius, slope, circumference, area, perimeter, and angles. Such programs may also perform reflections, dilations, rotations, and various constructions.

Coordinate systems can extend students' understanding of properties of individual shapes or classes of shapes by helping develop the notion of *congruence* among shapes. First, ask the students to express their own definition of *congruence*. From their responses, emphasize the ideas that are consistent with the notion that congruent figures have the same shape and size. To reinforce this idea, give the students the coordinates of a rectangle that has vertical and horizontal sides, and ask them to draw the figure on a coordinate grid, as depicted in figure 2.1. Using the coordinates of the rectangle, the students should easily be able to determine the lengths of the sides and the area of the figure. Have the students label the vertices of their figures with letters.

Fig. **2.1.**

A rectangle drawn on a coordinate grid

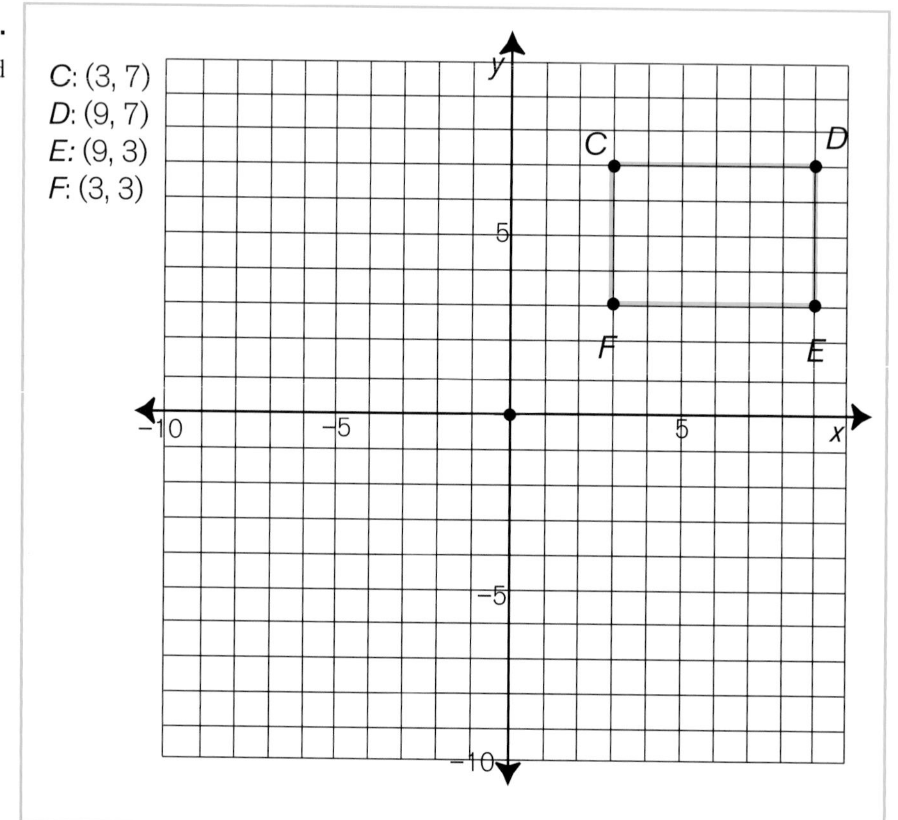

"Stitching Quilts into Coordinate Geometry" (Westegaard 1998) can be used with students who are familiar with coordinates, slope, and equations for lines. The article, which can be found on the CD-ROM, includes two activity sheets that connect geometry to important algebra concepts.

Next, give the coordinates of a rectangle that has the same area as the first one but different side lengths. Ask, "Are these figures congruent? Why or why not?" Repeat this process with another type of figure. Then specify the coordinates of a triangle with sides of different lengths, and have the students draw it and label the vertices (see fig. 2.2). Ask the students to draw a second triangle that is congruent to the first. The students may use units such as the diagonals of the squares on the grid to make arguments about the congruence of their triangles. Ask the students to identify the corresponding angles and sides of the two figures.

As a closing exercise, present the students with two figures that are of the same class but not congruent, and ask them to change the figures to make them congruent. Have the students specify the coordinates of their new, congruent figures and share the process they used to ensure

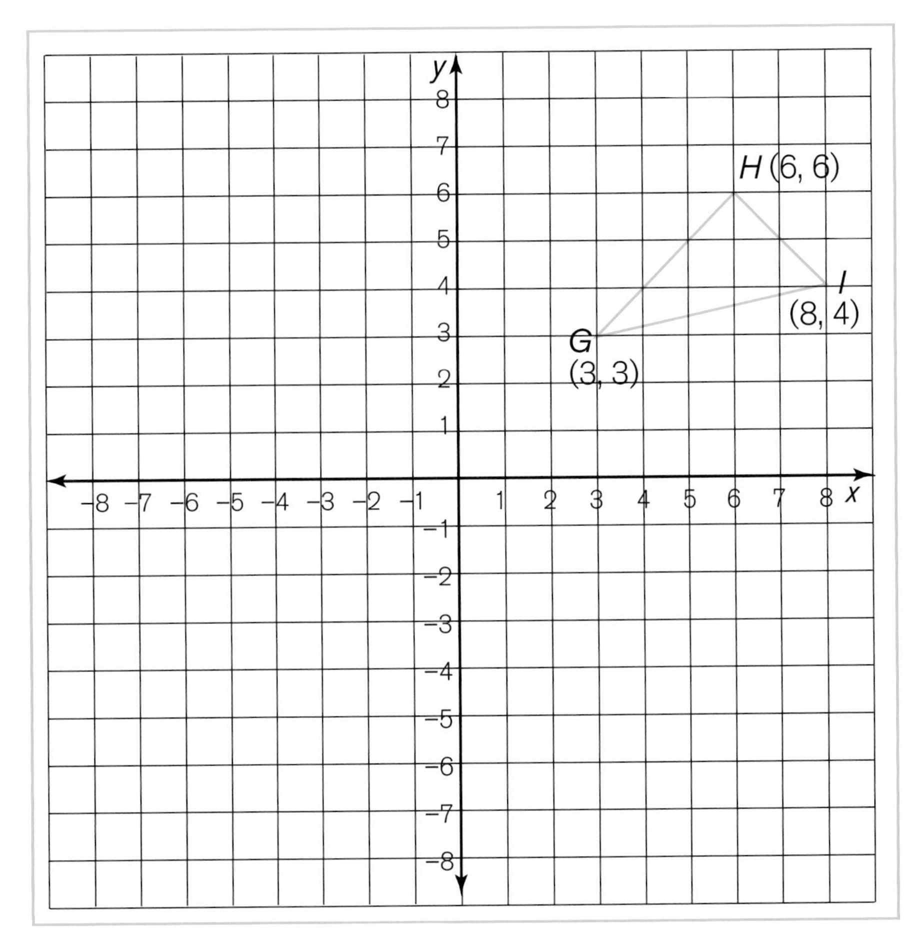

Fig. **2.2.**

A triangle drawn on a coordinate grid

that the two figures are indeed congruent. The following questions may help them verbalize their method:

1. How did you decide the two figures you created were congruent?
2. How could you use the coordinate plane to convince someone that your figures are congruent?

Many students may create figures that "appear" congruent but are not. They may not have studied distance and slope in coordinate planes and may therefore be unable to use the coordinate plane to establish the congruence of shapes.

Constructing Geometric Figures in Coordinate Space

Goals

- Reinforce or develop graphing skills
- Explore properties of shapes in a coordinate system

Materials

p. 96

- A copy of the blackline master "Constructing Geometric Figures in Coordinate Space"
- Grid paper or the Coordinate Geometry Tool (both available on the CD-ROM)
- Rulers

Activity

This activity can also be done in conjunction with the Coordinate Geometry Tool applet on the CD-ROM.

The students use grid paper to construct the figures described in the left-hand column of the table on the activity sheet and list the coordinates of the figures in the right-hand column. For shapes 1–13, encourage the students to use integers as the coordinates of the vertices of the shapes they create.

Discussion

Using grid paper to plot the coordinates of geometric figures helps students understand important properties of polygons and relationships among the properties while also developing their skills with coordinate systems. The exercise allows the students to plot coordinates that correspond to various descriptions of figures and then reflect on the characteristics of those figures. Of course, the descriptions in the table are only a starting point for teachers who may wish to pursue this activity further. Many of the descriptions in the table can be satisfied by more than one construction. It is important, therefore, that students share their different representations with peers, justify their solutions, and discuss what makes their solutions unique or similar.

In addition to the previous explorations, students should be given other opportunities to think explicitly about the ways in which space is represented in coordinate systems. Understanding representations on coordinate systems is particularly important as students encounter graphical representations of functions in algebra. The following activity requires students to look for patterns in the coordinate plane. Doing so not only contributes to the development of greater spatial awareness but also begins to lay the geometric foundation essential to understanding graphical representations of algebraic statements.

Exploring Lines, Midpoints, and Triangles Using Coordinate Geometry

Goals

- Recognize relationships among points on a coordinate plane
- Locate the midpoint between given points

Materials and Equipment

- A copy of the blackline master "Exploring Lines, Midpoints, and Triangles Using Coordinate Geometry" for each student
- Rulers

p. 97

Activity

This activity consists of three sections: In Graphs, Coordinates, and Lines, the students plot points and study the commonalities and differences among them. In Graphs, Coordinates, and Midpoints, the students draw line segments, find their midpoints, and explore the relationship between the coordinates of the midpoints of the segments and the coordinates of their end points. Graphs, Coordinates, and Triangles gives the students the coordinates of the vertices of three congruent triangles and asks them to draw the triangles and explore how the coordinates of the vertices are related. Students then repeat the activity with the coordinates of three similar triangles to discover a different relationship.

This activity can also be done in conjunction with the Coordinate Geometry Tool applet on the CD-ROM.

Similarity and the Coordinate Plane

The coordinate plane may also be used to develop students' understanding of similarity. Building on the scaling exercises in chapter 1, the next activity, Similarity and the Coordinate Plane (adapted from O'Daffer and Clemens [1992, p. 239]), explores dilations as a method for creating similar figures in coordinate space. As they engage in scaling, students will also be reinforcing previously developed skills involving the coordinate plane.

Similarity and the Coordinate Plane

Goals

- Explore dilations as a method for creating similar figures in coordinate space
- Reinforce skills involving the coordinate plane

p. 100

Materials

- Grid paper (available on the CD-ROM)
- A copy of the blackline master "Similarity and the Coordinate Plane" for each student
- Rulers

Activity

To reinforce the students' skills in representing figures in the coordinate plane, have them draw on grid paper the irregular pentagon and the triangle in figure 2.3, or copy the diagram for them. Instruct the students to identify the coordinates of each of the vertices of the two figures. Next, ask them to multiply the coordinates of points A, B, and C by 2, plot these new points on the grid, and label them A', B', and C', respectively. Have them do likewise for pentagon $DEFGH$ but multiply the coordinates of each point by 1/2 and label the new figure $D'E'F'G'H'$. (See fig. 2.4.) Discuss with the students the questions on the activity sheet.

Fig. **2.3.**

(a) A pentagon and a triangle drawn on grid paper and (b) the original pentagon and triangle and similar figures drawn on grid paper

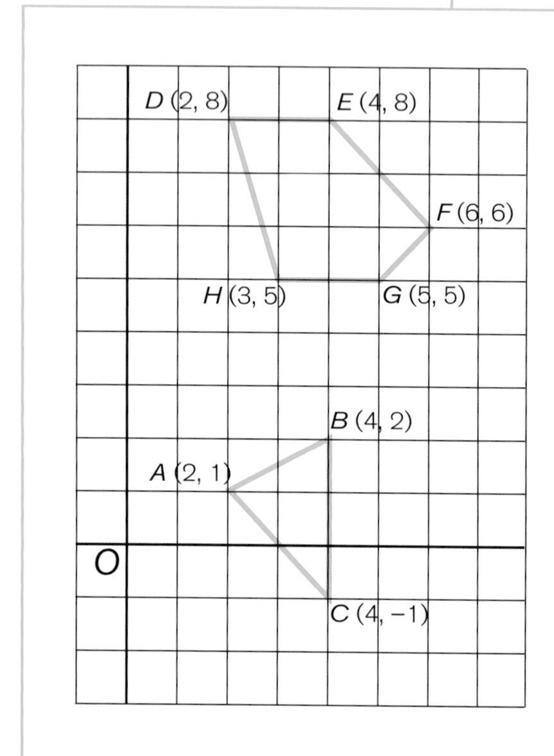

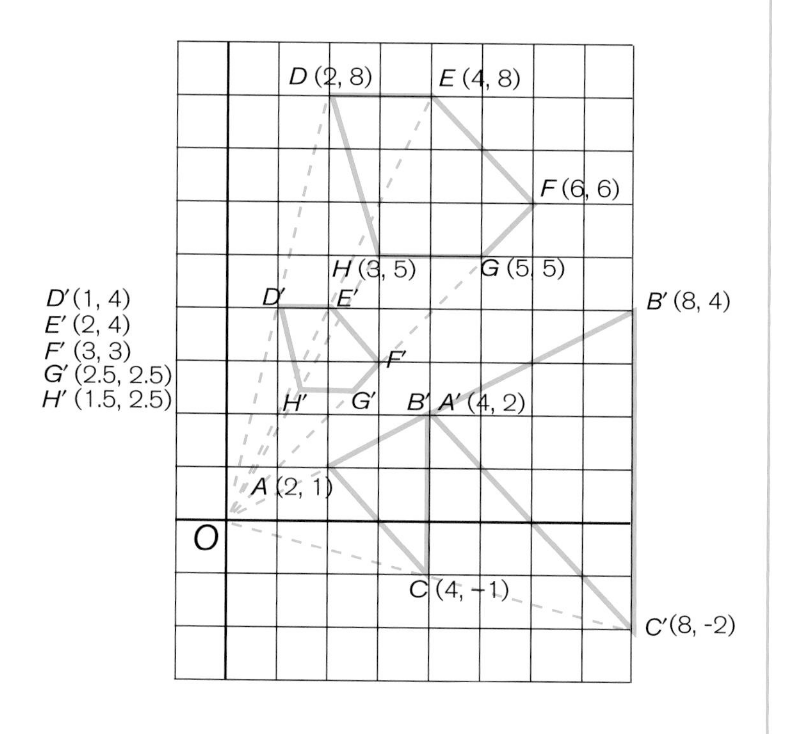

Discussion

This activity gives students an opportunity to explore further the similarity of figures. It is important in the discussions to make sure students understand the difference between similarity and congruence. If two figures are congruent, they match each other exactly (e.g., two rectangles having the same length and width). Similarity goes beyond the relationship of congruence (congruence may be viewed as a special case of similarity). In general, two figures are similar if they have the same shape (e.g., two squares) but not necessarily the same size (e.g., one square with sides of length 4, a second square with sides of length 8). The previous activity reinforces these ideas as students construct similar polygons by multiplying the coordinates by different numbers. You could also use this exercise to discuss scaling, by drawing attention to the different sizes of the scaled figures compared with the size of the original.

This activity can also be done in conjunction with the Coordinate Geometry Tool applet on the CD-ROM.

Extensions

As students continue working with the coordinate system, they can extend their investigations beyond basic geometric concepts by connecting the concepts to algebraic ideas—slope, for instance. As a means of introducing students to the slope of a line, have them do some basic movements in coordinate space. They might start at a given point—say, (0, 0)—and move right 1 and up 1, plotting a point after each set of moves. Repeat this process, perhaps starting from the same point but going right 1 and up 3. Have the students connect the points and then plot two more points and connect them. Lead a class discussion about the difference in the "steepness" of the two lines. Asking the students to describe how their movements led to the changes in slope serves as an intuitive introduction to the concept of slope and how it relates to movements within coordinate space.

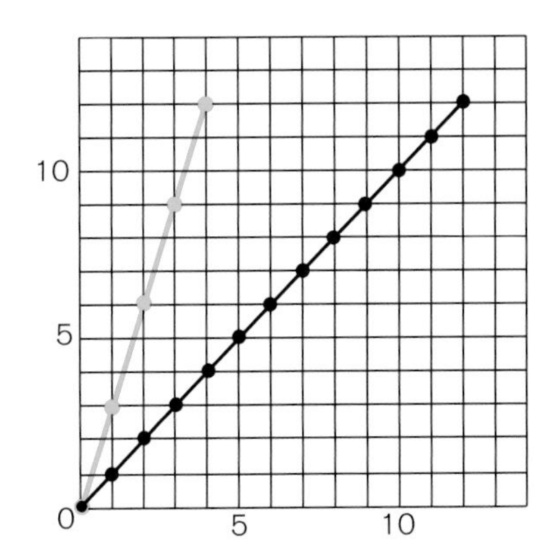

As an additional review or as an introduction, you might have the students start at (0, 0) and then move right 1 and down 2. Repeat the exercise several times, using different coordinates for the second point, and have the students connect pairs of points. This task sets the stage for a discussion about the *direction* of the segments, that is, positive and negative slopes. The students should also discuss the slopes of horizontal and vertical lines. They might be led to understand these special cases through induction. They can informally explore the slopes of lines that are gradually getting steeper and those that are gradually getting flatter. Ask the students to make a guess about what a line with a slope of 0 would look like, then ask them what the slope of a vertical line might be. To solidify these concepts, direct the students to construct a figure with at least one side with a slope of 0 or at least one side with the slope undefined. Discuss students' answers. What is the most common figure? The least common? These intuitive explorations will help the students as they further explore properties of polygons.

Triangles are helpful in examining the formula for determining the slope of a line. Have the students draw the triangle represented in figure 2.4 on their own graph paper. After the students identify and label the vertices of the triangle, begin an exploration of the slope by considering the relative "steepness" and direction (positive or negative) of line segments *AB* and *BC*. Encourage the students to think about a way to

Fig. **2.4.**

A triangle with vertex A at (0, 0), B at (3, 4), and C at (6, 0)

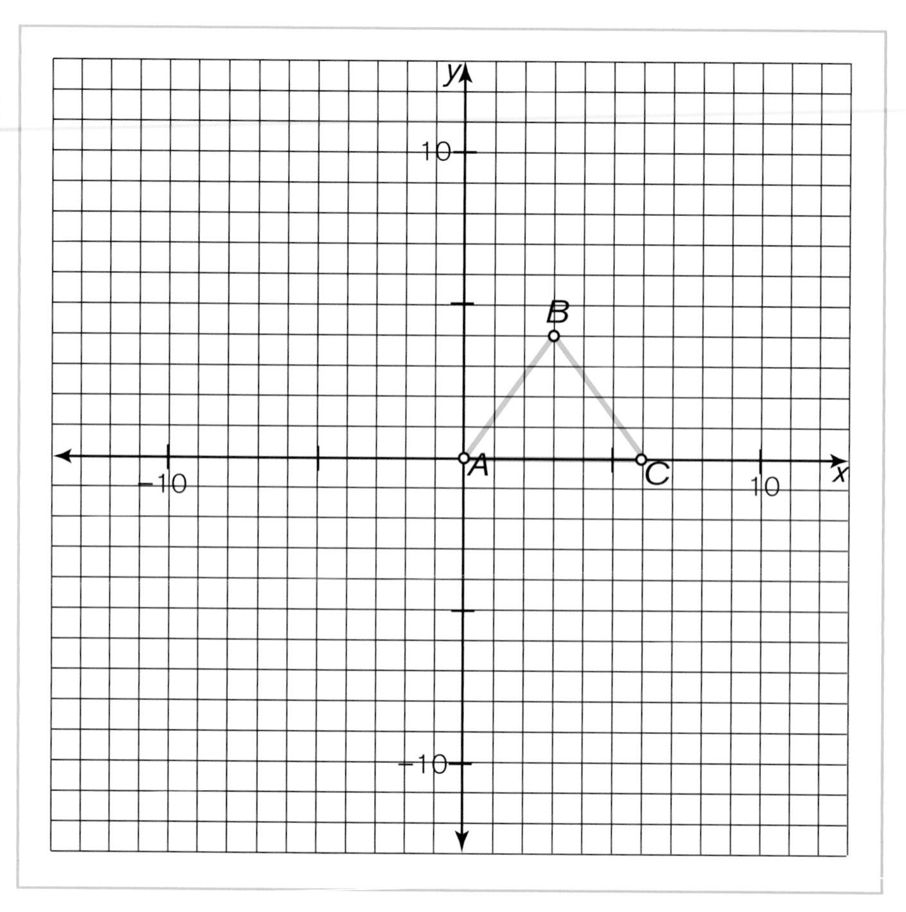

describe the slope of a line by starting with the base (in this example, segment AC) of the triangle. It is important to start with the base because the slope of segment AC is 0. One way to help the students recall that the slope is 0 is to illustrate that segment AC has no vertical rise. That is, the y-coordinates are the same for both point A and point C. With some guided discussion, the students may begin to think about slope as the relationship between the coordinates of two corresponding points. For example, to move from A to B, the students must go right 3 and up 4. Some students benefit from considering the movement along the y-axis first and the movement along the x-axis second. They may then be prompted to discover the following formula:

$$\text{Slope} = \frac{\text{vertical change (rise)}}{\text{horizontal change (run)}} = \frac{4}{3}$$

Have the students find the slope of the remaining side of the triangle. They should arrive at – (4/3). Discuss why sides AB and BC have slopes that are "opposites." Ask, "What kind of triangle have we constructed?" (isosceles) "How do you know?"

The following activity builds on this initial exposure to slope by having students examine various figures and make conjectures about their properties and relationships to one another. The intent of the activity is to encourage students to determine the slopes of lines and to use the slopes to examine figures—particularly those that are composed of parallel and perpendicular lines.

Exploring the Slopes of Parallel and Perpendicular Lines

Goals

- Develop an understanding of slope
- Compute and compare the slopes of line segments
- Explore characteristics of parallel and perpendicular lines

Materials and Equipment

p. 101

- A copy of the blackline master "Exploring the Slopes of Parallel and Perpendicular Lines" for each student
- Rulers
- Grid paper (available on the CD-ROM)
- Protractors

Activity

First, focus the students' attention on the three shapes drawn on the grid at the top of the activity sheet. Have the students determine the slope of each of the sides of the polygons by asking, "What is the rise? What is the run?" For example, in trapezoid $JKLM$, L is "right 4 and up 6" from M, so the slope is 6/4, or 3/2, and K is "right 2 and up 3" from J, so the slope is 3/2. After the students have completed the activity sheet, follow up with a discussion.

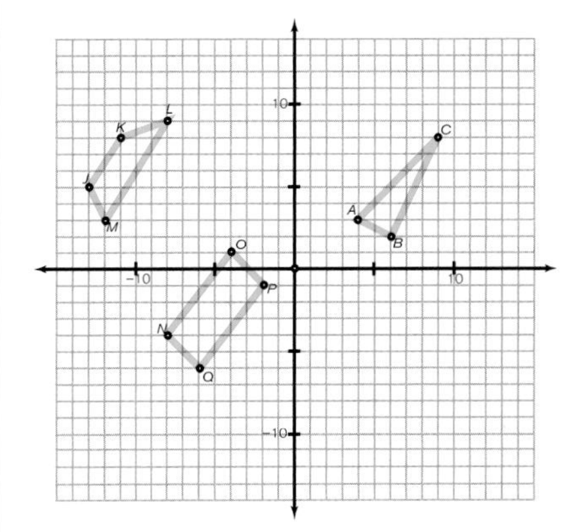

Discussion

Beginning with trapezoid $JKLM$, compare the slopes of each of the line segments that form the figure. Ask,

- "Which slopes are the same?" (The slopes of JK and ML are both 3/2.)
- "In a figure, what is the relationship between sides that have the same slope?" (They are parallel.)

The same questions may be asked about parallelogram $NOPQ$. Also ask, "What is the primary difference between $JKLM$ and $NOPQ$?" (The parallelogram has two pairs of parallel sides.) Encourage the students to use a protractor to confirm that $NOPQ$ is not a rectangle.

Next, direct the students' attention to the slopes of the sides of triangle ABC. Although no sides of this figure have the same slope, as in the other figures, there is a notable relationship between the slopes of segments AB (–1/2) and BC (2). Ask the students to describe the relationship between the slopes (they are negative reciprocals), as well as the relationship between line segments AB and BC (they are perpendicular). In a discussion, lead the students to conjecture about the slopes of parallel and perpendicular lines. Have them test these conjectures by creating several additional figures that have either right angles or parallel sides and then determining whether these examples confirm their hypotheses.

To reinforce these concepts further, give the students two points and ask them to use the two points as vertices as they construct the following figures: (1) a figure with one pair of parallel sides, (2) a figure with two pairs of parallel sides, (3) a figure with one pair of perpendicular sides, and (4) a figure with two pairs of perpendicular sides. Have the students share their drawings with partners to have the partners confirm that the constructions meet the criteria. Throughout the discussion, be mindful of helping the students think deeply about the ways in which slopes can be used to help them understand various shapes and their properties.

Conclusion

The explorations in this chapter afford students opportunities to develop their spatial skills using coordinate representations. Many of the concepts in this chapter form a foundation for studying important ideas in algebra. The approaches emphasized throughout the chapter require students to explore representational systems and also to explain and describe the relevant geometric principles. The skills developed in the activities are important as students explore symmetry and transformations as well as construct geometric models to solve problems. The activities have emphasized important goals from *Principles and Standards for School Mathematics* (NCTM 2000). These goals include developing students' ability to use coordinate geometry to represent and examine the properties of geometric shapes, including figures with pairs of parallel or perpendicular sides. Coordinate geometry serves as a context for describing sizes, positions, and orientations of shapes and for extending ideas about congruence and similarity. Transformations in the coordinate plane are presented in the next chapter. Experiences with the kinds of tasks presented in this chapter help students extend their abilities to use various tools for representation and analysis.

NAVIGATING *through* GEOMETRY

GRADES 6–8

Chapter 3
Transformations and Symmetry

Important Mathematical Ideas

Students can trace one triangle on tracing paper and physically translate, reflect, or rotate it to match another triangle. This inductive experiment to determine coincidence has a parallel deductive proof that the two triangles are congruent.

Motion plays a significant role in our daily life. Students can relate motions—reflections, translations, and rotations—to their everyday experiences of riding a bike, running on the playground, or going to a department store. Hands-on explorations of motion are appealing to middle-grades students who are making the transition from generalizing from inductive, concrete experiences to making more deductive, abstract generalizations.

Objects that are translated, reflected, or rotated preserve their size and shape, although their location and orientation may change. This chapter focuses on ways transformations help students understand similarity and symmetry and help them classify polygons, thus extending the work begun in chapter 1 on recognizing the properties and characteristics of shapes.

Principles and Standards for School Mathematics (National Council of Teachers of Mathematics [NCTM] 2000) indicates that middle-grades students should have experiences that develop the skills and understandings that allow them to successfully—

- describe sizes, positions, and orientations of shapes under informal transformations such as flips, turns, slides, and scaling;
- examine the congruence, similarity, and line or rotational symmetry of objects using transformations. (P. 232)

"Transformational geometry offers another lens through which to investigate and interpret geometric objects." *(NCTM 2000, p. 235)*

What Might Students Already Know about These Ideas?

Students' prior experiences with transformations may have included reflections, translations, and rotations (flips, slides, and turns). They may have determined congruence by placing one figure on top of another and extended that strategy to turning and reorienting figures to demonstrate congruence. They are probably able to identify a line of symmetry and find the lines of symmetry in figures.

The orientation of a figure in a plane may, however, distract some students from identifying the shape. Younger students are especially distracted by a geometric figure that is not oriented so that its base is "at the bottom" and parallel to the edge of the page. Even some middle-grades students may fail to recognize a square when it is displayed as shown in figure 3.1. Students frequently refer to this shape as a "diamond."

Fig. **3.1.**

Some students may call a square in this orientation a "diamond."

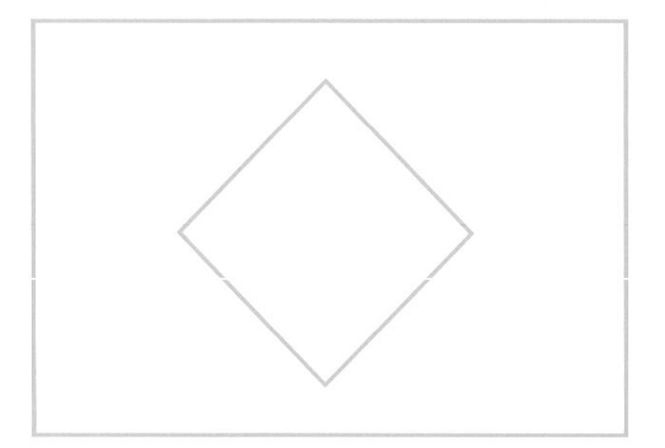

Students who have had limited experience with transformations and other spatial-reasoning tasks often identify figures according to their appearance. They are influenced by irrelevant attributes of a figure, such as its orientation, and are unlikely to recognize a rotated figure. With more experience, students analyze a figure by examining its parts and the relationships among them.

The following activity allows teachers to preassess students' ideas about transformations. It focuses on reflections in the plane, reorientations that do not change any of the properties of the figure. Transformations that preserve congruence are called *rigid motions.*

Reflection of Images

Goals

To assess students'—

- understanding that reflections preserve the properties of figures;
- ability to draw reflections of figures.

Materials and Equipment

p. 103

- A copy of the blackline master "Reflection of Images" for each student
- Rulers
- Tracing paper

Activity

Students draw the image of each of four figures after the figures have been reflected over the y-axis of a coordinate grid, and then they draw the image of each figure after it has been reflected over the x-axis.

Discussion

The students might verify their answers by tracing the figure and then turning the paper over to show the reflected image of the original. If they say that the figures are the "same," ask them to explain in what ways the figures are the same. They should mention the lengths of the sides and the measures of the angles. Next, ask them why these properties remain unchanged. They should recognize that a reflection merely reorients an object without changing anything else about it. If the students fail to notice that the shape and size have not changed, prompt them to reflect the figure over one of the axes and observe the results. Once again, focus the students' attention on the figure's side lengths and angles. The degree to which students are troubled by the orientation of a figure can help you distinguish students with limited prior experiences from those who are familiar with transformations. For students with some background in transformations, orientation will be less troubling, and they should be able to use appropriate vocabulary to describe transformations.

Selected Instructional Activities

Tracing tasks give students experience in determining the image of a figure under a transformation. Exploring relationships between the preimage (original figure) and the image in translations, reflections, and rotations is the goal of the following activity (adapted from the Reconceptualizing Mathematics Project n.d.). The students may be more familiar with the terms *slide*, *flip*, and *turn* than with *translation*, *reflection*, and *rotation*. Allow the students to use the familiar terms to describe their actions, but also introduce the geometric vocabulary for such motions.

The CD-ROM includes an interactive applet, Transformation Tools, that can help students visualize transformations.

Translations, Reflections, and Rotations

"Transformations can become an object of study in their own right. Teachers can ask students to visualize and describe the relationship among lines of reflection, centers of rotation, and the positions of preimages and images."
(NCTM 2000, p. 236)

p. 104

Goals

- Explore relationships between the preimage and the image in rigid motions
- Develop appropriate language to describe rigid motions
- Perform three rigid transformations: reflections, translations, and rotations

Materials and Equipment

- A copy of the blackline master "Translations, Reflections, and Rotations" for each student
- Tracing paper (waxed paper or any paper that is easy to see through)
- Rulers

Activity

The students perform a translation, a reflection, and a rotation according to the instructions on the activity sheet. They should be encouraged to execute each transformation carefully so that the characteristics of each figure are preserved. The students should repeat each transformation with a shape of their choice to give them practice in performing transformations.

Translation

Translations are described by answering the questions, How far? and In what direction? Translations are represented by a translation arrow, or *vector*, that indicates how far to translate and in what direction. The students are instructed to use a vector to complete the translation.

Some people prefer to use the terms *original figure* and *image* to distinguish between the starting shape and the ending shape.

Reflection

The line that is the axis of the reflection is sometimes called the *line of reflection* or the *mirror line*. Corresponding points on the reflection (*image*) and the initial figure (*preimage*) are the same distance from the line of reflection. The students are instructed to specify a line of reflection over which they flip the shape they designate.

Is reflection commutative? A figure is reflected first across the *x*-axis and then across the *y*-axis. Is the resulting image congruent to, and in the same orientation as, the image formed if the figure were reflected first across the *y*-axis and then across the *x*-axis?

Rotation

To describe a rotation, it is necessary to specify the point on which the rotation is centered, called the *center of rotation*, and an angle that shows the size and direction (clockwise or counterclockwise) of the rotation. The students are instructed to specify both a center of rotation and an angle of rotation for this transformation.

Discussion

Ask the students to observe the preimage (the original figure) and the image. The discussion should focus on the characteristics of both figures and the characteristics of the motion itself. Ask questions such as

How did the figure move? and Can you describe the path of the motion? In observing the figures, questions such as What changed? and What stayed the same? will help the students analyze the relationship between the preimage and the image. The motion of a reflection is like a flip; the resulting image is congruent with the preimage, but the orientation has changed. A translation is a slide; it always moves in a straight line. It may be in any one direction. The resulting image is congruent with the preimage, but its position has changed. A rotation is a circular motion about a fixed point, which may or may not be a point on the original figure. The resulting figure is congruent with the preimage, but its position has changed.

From their sketches, the students should be able to tell how the orientations of the original figure and its image are related for each type of rigid motion. *Orientation* refers to the ordering of the points in a figure either clockwise or counterclockwise. If, for example, you pick three points on a figure and assign them *P*, *Q*, *R*, then the order *P-Q-R* gives a clockwise or counterclockwise orientation to the figure.

Students who have identified the basic characteristics of rigid motions can go on to more-analytical observations. They can, for instance, connect corresponding points of the original figures and their reflection images and observe that the segments that connect the corresponding points are perpendicular to the line of reflection and that the line of reflection bisects the segments connecting the corresponding points. They can draw and test many vectors that result in the same translation and observe that all the vectors have the same lengths and lie parallel to one another. The students can examine rotations and discover that the angle between the "starting ray" and the second ray determines the size of the turn and that, in order to achieve the same degree of rotation, points farther from the center of rotation have to "move farther" along an arc than points closer to the center do.

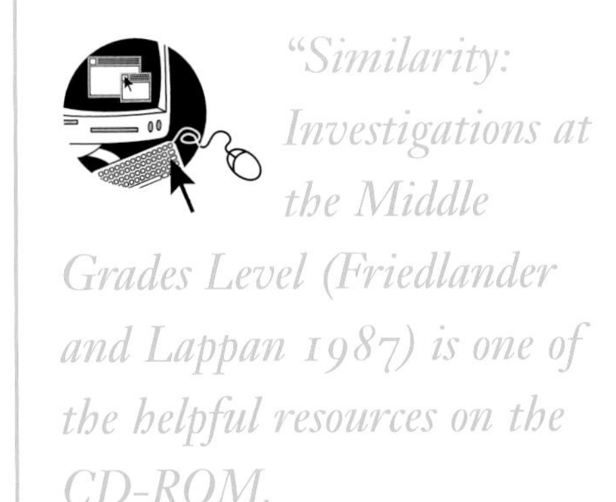

"Similarity: Investigations at the Middle Grades Level (Friedlander and Lappan 1987) is one of the helpful resources on the CD-ROM.

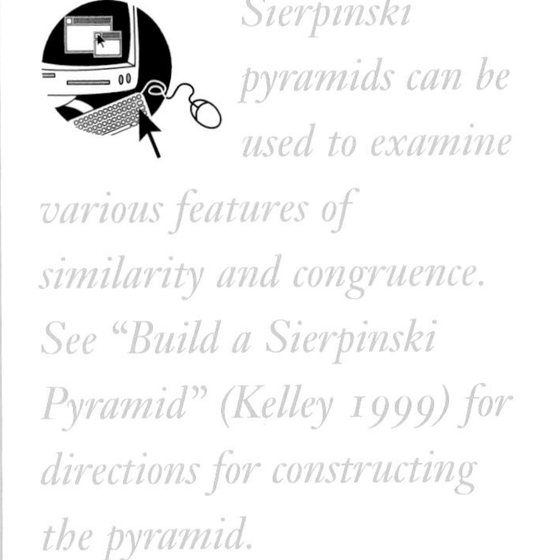

Sierpinski pyramids can be used to examine various features of similarity and congruence. See "Build a Sierpinski Pyramid" (Kelley 1999) for directions for constructing the pyramid.

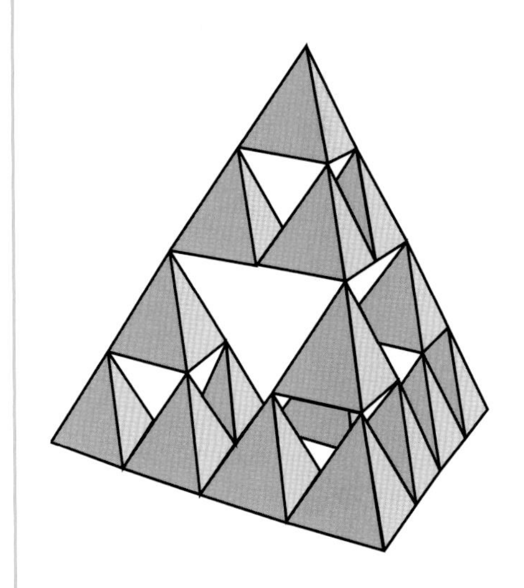

Examining Congruence and Similarity

The previous chapters have presented several activities that involve similarity and congruence. Chapter 1 discussed some properties of similar and congruent figures and introduced dilations. Chapter 2 related those concepts to coordinate geometry. You may want to review some of the ideas in the first two chapters before continuing your class's explorations with related concepts.

Using Scale Factors

Goals

- Draw similar figures
- Use scale factors
- Determine the relationship between corresponding parts of similar figures

Materials and Equipment

p. 107

- A copy of the blackline master "Using Scale Factors" for each student
- Scissors
- Rulers
- Protractors

Activity

Given a projection point and a scale factor for each figure, the students draw a scaled figure for a triangle and a rectangle. The method is similar to that used by Sheri to dilate her figure (see the solution for "Dilating Figures," in chapter 1): From the projection point, draw rays through the vertices of the figure. Place the vertices of the scaled figure at the points whose distances from P are the lengths of the segment from P to the corresponding vertices of the original figures multiplied by the scale factor.

Discussion

The students should be able to find the ratios of the corresponding sides of both the triangles and the rectangles. Ask questions like the following to guide the students' thinking:

- How are the ratios related to the size of the image? Specifically, how does the ratio relate to the perimeter of the image? (The perimeter increases or decreases by the scale factor.)
- How does the ratio relate to the area of the image? (The area of the image is the area of the original figure multiplied by the square of the scale factor.)

The students should also explore the relationship between the corresponding angles of the preimage and the image. They can cut out the figures and compare the sizes of the corresponding angles, or they can use protractors to measure the angles. It is important that the discussion focus on helping students understand what scale factors are, how scaled figures can be constructed using a projection point, and the relationships between the corresponding sides of similar figures and between the corresponding angles. In particular, the students should be able to distinguish between figures that are similar and figures that are congruent.

Broadening Students' Experience with Symmetry

Most middle-grades students recognize line symmetry in a figure, particularly when the line of symmetry is parallel to the longer side of the page. They have more difficulty in determining whether a figure has rotational symmetry. Several reasons may be given for this difference:

1. Whereas line symmetry is commonly addressed in elementary-level mathematics materials, such materials include few examples of rotational symmetry.
2. Students find it more difficult to rotate objects in their minds than to reflect objects mentally (Clements and Battista 1992).
3. The most common model for line symmetry is a figure cut in half, not a figure reflected over a line so that the image coincides with the preimage. Without the latter model, students' ideas of reflection symmetry may be static, making it difficult for them to understand how a movement (rotation) can demonstrate symmetry.

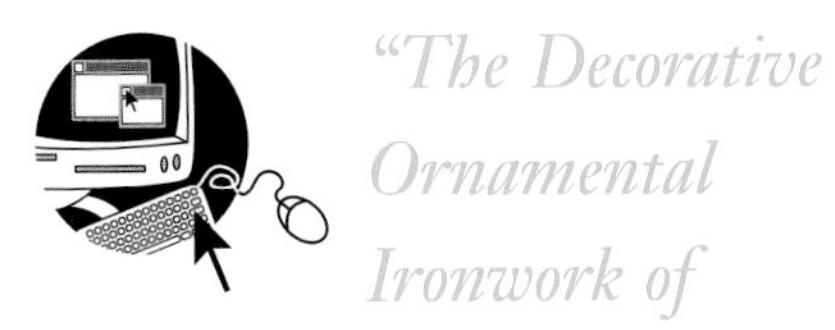

"The Decorative Ornamental Ironwork of New Orleans: Connections to Geometry and Haiti" (Germain-McCarthy 1999), available on the CD-ROM, includes activities involving symmetry in a real-world context.

Explorations with Lines of Symmetry

Goals

- Explore symmetry of figures
- Determine whether a line is a line of symmetry and identify lines of symmetry

Materials and Equipment

p. 108

- A copy of the blackline master "Explorations with Lines of Symmetry" for each student
- Rulers

Activity

For each of four figures with dashed lines drawn through them, the students determine whether the lines represent lines of symmetry, and they explain the reasons for their answers. They also draw figures that have line symmetry and locate the lines of symmetry.

"Above and beyond AAA: The Similarity and Congruence of Polygons" (Slavit 1998) offers important ideas and activities related to the development of an intuitive understanding of these concepts.

Discussion

The students may have had some experience with symmetry. The activity can give you a good indication of what they may already know about the topic. You can extend the task by asking them to find other (or all) lines of symmetry in each figure. You can also extend the discussion to include regular polygons. Have the students find the lines of symmetry in an equilateral triangle, a square, a regular pentagon, and a regular hexagon. For each polygon, ask questions such as these:

- How many lines of symmetry does the figure have?
- What is the relationship between the number of sides in a regular polygon and the number of lines of symmetry? (They are equal.)
- Does your conjecture work with a regular octagon and a regular decagon?

The students may not recognize that lines of symmetry can pass through the midpoints of opposite parallel sides, through opposite vertices, or through a vertex and the midpoint of the opposite side. This task can help students make generalizations relating the number of sides in a regular polygon to the number of its lines of symmetry. Regular polygons can also serve as examples of rotational symmetry.

Extensions

Students' thinking can also be extended by the following tasks. The students should be able to justify their thinking either verbally or in writing. As they draw figures to illustrate examples and counterexamples, their geometric reasoning will be broadened.

> Tell whether each statement is always true (true), sometimes true (maybe), or never true (false) when applied to polygons. Support your responses with figures.

- A line of symmetry passes through two vertices. (maybe)
- A line of symmetry passes through at least one vertex. (maybe)
- A line of symmetry divides the figure into triangles. (maybe)
- A line of symmetry divides the area in half. (true)
- Every polygon has at least one line of symmetry. (false)
- Any two lines of symmetry bisect each other. (maybe)

Fold a square on the diagonal, and respond true or false to the following statements:

- The two halves match up.
- The two halves are congruent.
- The two halves have the same area.
- One half is the mirror image of the other.
- The two halves are isosceles right triangles.

All these statements are true and help to demonstrate that a diagonal of a square is a line of symmetry.

Understanding Types of Symmetry

Goals

- Explore reflection and rotational symmetry

Materials and Equipment

pp. 109, 110

- Two transparency copies of the blackline master "Alphabet Symmetry"
- A transparency copy of the blackline master "Square Symmetry," with the two squares cut apart
- An overhead projector

Activities

Exploring Symmetry with the Alphabet

Display on the overhead projector a transparency of "Alphabet Symmetry." The second copy of the transparency can be used as an overlay with which to demonstrate symmetry in the letters. Present the following problem to the students:

> Vinh and Maria were looking at the capital letters of the alphabet to decide which ones have symmetry. Vinh said, "I think the letters A, M, and E have symmetry." Maria agreed with him but added, "I think that the letters S and Z also have symmetry."

Ask the following questions to guide students' thinking:

- How is Vinh deciding which letters have symmetry? (He appears to be thinking about reflections.)
- What other capital letters would he say had symmetry? (B, C, D, H, I, O, T, U, V, W, X, and Y have line symmetry in the typeface shown.)
- How is Maria deciding which letters have symmetry? (She seems to be considering rotations as well as reflections.)
- What other capital letters would she say had rotational symmetry? (H, I, N, O, X also have rotational symmetry in this typeface.)

Discussion

Students who are familiar with rigid motions will have little difficulty determining which capital letters have reflection symmetry, so they should be able to name the remaining letters that have line symmetry. Maria's adding S and Z to the list of letters having symmetry pushes students to think about symmetry in a different way. If no students raise the possibility of using a rotation to determine symmetry, ask, "What other motion could Maria be using to test for symmetry?"

Rotational Symmetries of a Square

Display on the overhead projector the two transparencies cut from "Square Symmetry." Overlay the two squares with point A coinciding with point A' and point B coinciding with point B'. The green square represents the original figure. Then rotate square $A'B'C'D'$ counter-clockwise about the center point until it reaches its original position, noting the number of times the squares align during a full 360-degree rotation.

Discussion

A figure that has rotational symmetry can be rotated about a fixed point until it fits the original space it occupied. This fixed point is referred to as the *center of rotation*. Square $ABCD$ is said to have rotational symmetry of 90 degrees. Ask the students why they think such terminology is appropriate. Have them explore whether the square also has rotational symmetries of 180 degrees and 270 degrees. Although every figure has 360-degree rotational symmetry, a 360-degree rotation is considered only when a figure displays other rotational symmetries. So a square is also said to have fourfold rotational symmetry (see fig. 3.2). Have the students verify that clockwise rotations do not reveal any new rotational symmetries in the square.

Fig. **3.2.**

Fourfold rotational symmetry of a square

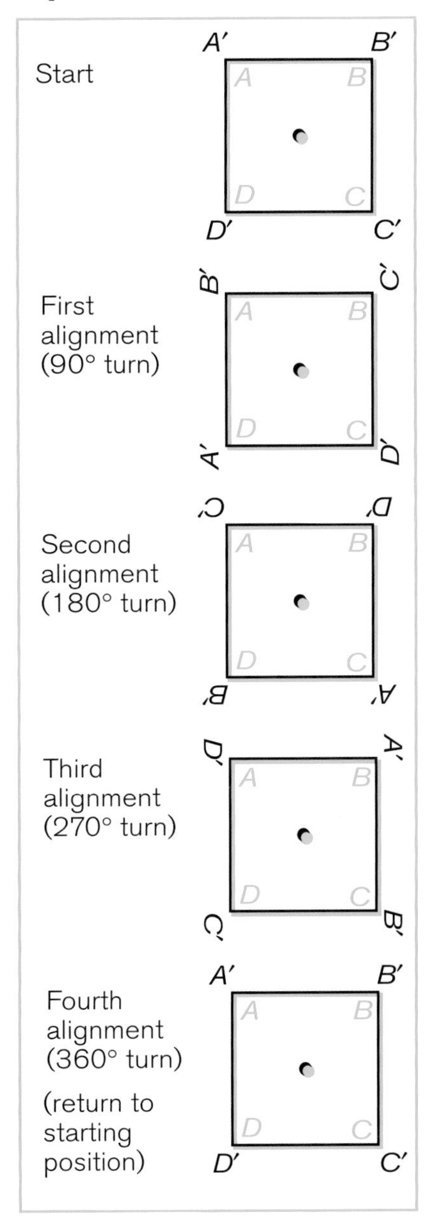

Rotational Symmetry and Regular Polygons

Goals

- Develop skills with rotational motion
- Identify rotational symmetry
- Identify the number of rotational symmetries of regular polygons
- Formulate and test conjectures about geometric relationships
- Represent geometric relationships using algebra

Materials

p. 111

- A copy of the blackline master "Rotational Symmetry and Regular Polygons"
- Tracing paper
- Rulers or straightedges
- Patterns for regular polygons

Activity

The students draw a large equilateral triangle on one piece of tracing paper, locate the center of rotational symmetry by folding the bisectors of two of the angles, and label the point of intersection as the center of rotational symmetry. They place a mark outside the triangle to use as a reference point. They then copy the triangle, the center of rotation, and the reference point onto another piece of tracing paper and place the copy over the original, hold a pencil tip at the center of rotational symmetry, and slowly rotate the copy until its image overlays the original triangle. They observe the number of times the image coincides with the original in one complete (360°) rotation (three times). (See fig. 3.3.) The students repeat these steps with other regular polygons, record their data, look for patterns in their data, and make conjectures about the relationship between the number of sides of a regular polygon and the number of its rotational symmetries.

Fig. **3.3.**
A method for determining the number of rotational symmetries in an equilateral triangle

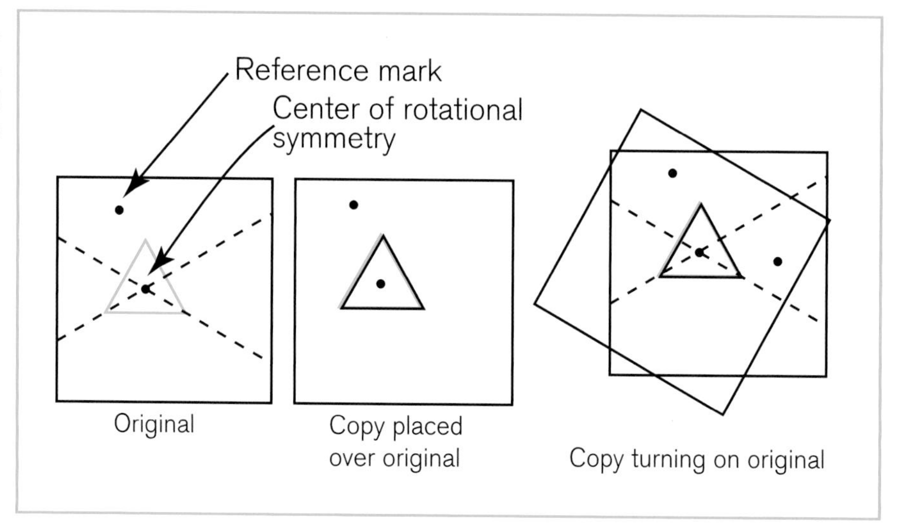

Discussion

The students should be encouraged to express their conjectures both in writing and algebraically. Throughout this exercise, monitor the students' ability to perform the rotations and to recognize when a figure has rotational symmetry. Suggest that the students pause when they find rotational symmetry to make a note so they can keep track of their information. They can then continue until they have completed the 360-degree rotation. As an extension, have the students try several figures that are not regular. Ask, "What do you notice? How could you explain the differences between the results for regular and nonregular polygons?"

Extending Experiences with Symmetry

Principles and Standards for School Mathematics (NCTM 2000) calls for students in the middle grades to explore and examine the characteristics of shapes using a variety of materials in order to define special types of figures precisely and identify relationships among types of figures. Transformation geometry offers an additional basis on which students can analyze figures and make the distinctions necessary for classification. Students who use transformations to classify figures can test their ideas by actually performing the motions. This strategy benefits students who may have difficulty remembering the many terms and characteristics used to define classes of shapes by giving them an alternative way to analyze shapes and understand the relationships among the classes of shapes. For example, when students use reflection and rotational symmetry to classify quadrilaterals, they can test their thinking by rotating or reflecting a figure to see if it has the requisite symmetry. They can use rotations to define parallelograms as quadrilaterals that have 180-degree rotational symmetry. This approach makes sense intuitively and mathematically: in order for pairs of opposite sides to coincide when rotated, they must be parallel. To assess their understanding, we can ask students if a square is a parallelogram. Present them with the two shapes in figure 3.4, and ask, "If you rotate the square 180 degrees using G as the center of rotation, does it have 180-degree rotational symmetry?" Rotational symmetry of 180 degrees is also known as *point symmetry*. Using the same transformation and V as the center of rotation, have them test if the trapezoid is a parallelogram.

When students have had experience with rigid motions and have used motions to explore reflection and rotational symmetry, they can apply symmetry to the sorting task Geodee's Sorting Scheme. (Refer to fig. 1.1 to spark ideas about how rotational symmetry can be used to classify shapes.) As the students discuss how they used symmetry to sort the quadrilaterals, help them make observations about the symmetry characteristics of various types of quadrilaterals. The discussion can lead to the development of a classification scheme.

If students are not already familiar with reflections, translations, and rotations, these exercises dealing with rigid motions will challenge their current level of geometric thinking.

Fig. **3.4.**

A square and a trapezoid

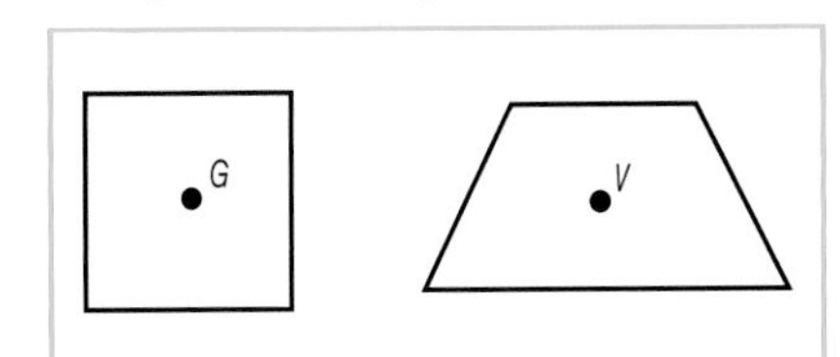

Drawing Figures with Symmetry

Goals

- Apply concepts of symmetry
- Recognize symmetry as an identifying property

Materials and Equipment

- Paper for drawing
- Chart paper for reporting

Activity

Have the students work in pairs or small groups to try to draw four-sided figures with the indicated number of lines of symmetry:

1. Exactly four lines of symmetry (square)
2. Exactly three lines of symmetry (none)
3. Exactly two lines of symmetry (rhombus, rectangle)
4. Exactly one line of symmetry (isosceles trapezoid, kite)
5. No lines of symmetry (scalene trapezoid; nonrectangular, nonisosceles parallelogram; irregular quadrilateral)

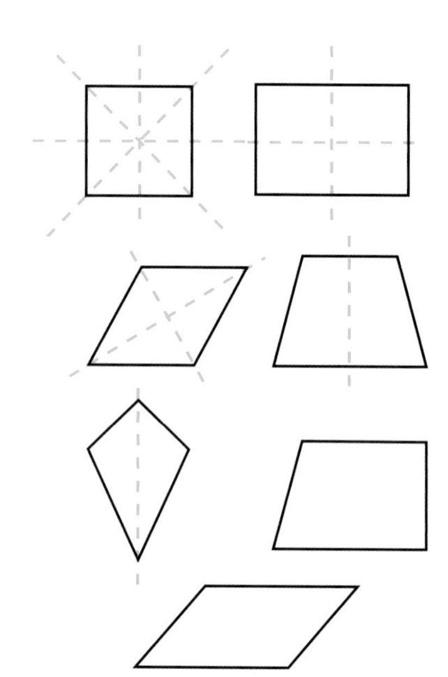

After the groups have completed their drawings, have them share them with the class.

Discussion

Explorations with symmetry help students understand that symmetry can be a means of classifying figures. Working only with four-sided figures helps students understand symmetry as an identifying property of a shape. As the students share their drawings, they will see that the conditions specified in each part of the task result in certain types of quadrilaterals. They will notice, for instance, that a square is the only quadrilateral with four lines of symmetry and that it is not possible to draw a quadrilateral with exactly three lines of symmetry. (If the students suggest that only an equilateral triangle can have exactly three lines of symmetry, you can build on their conjecture later when discussing regular polygons.) For a quadrilateral with exactly two lines of symmetry, most students draw a nonsquare rectangle, which has line symmetry through the midpoints of opposite sides. You might ask what other type of line symmetry a figure can have, thereby prompting the students to draw a rhombus, which has line symmetry through opposite vertices. Figures with exactly one line of symmetry include isosceles trapezoids and kites. Students often have to be prompted to include kites in their drawings. Many students identify the diagonals of a nonsquare rectangle as lines of symmetry because they divide the figure into two congruent parts. Folding the figure along the diagonals, however, will demonstrate that they are not lines of symmetry.

Extension

Extend the discussion of symmetry and promote a connection to classification by asking which of the figures also have 90-degree or 180-degree rotational symmetry. The students may discover that all figures with exactly two lines of symmetry also have 180-degree rotational symmetry and that those with exactly four lines of symmetry also have 90-degree rotational symmetry. Some students may discover that rotational symmetry also can be used to define a parallelogram. Figure 3.5 (from CRDG, Geometry Learning Project [n.d.]) shows how reflection and rotational symmetry can be used to build a classification hierarchy.

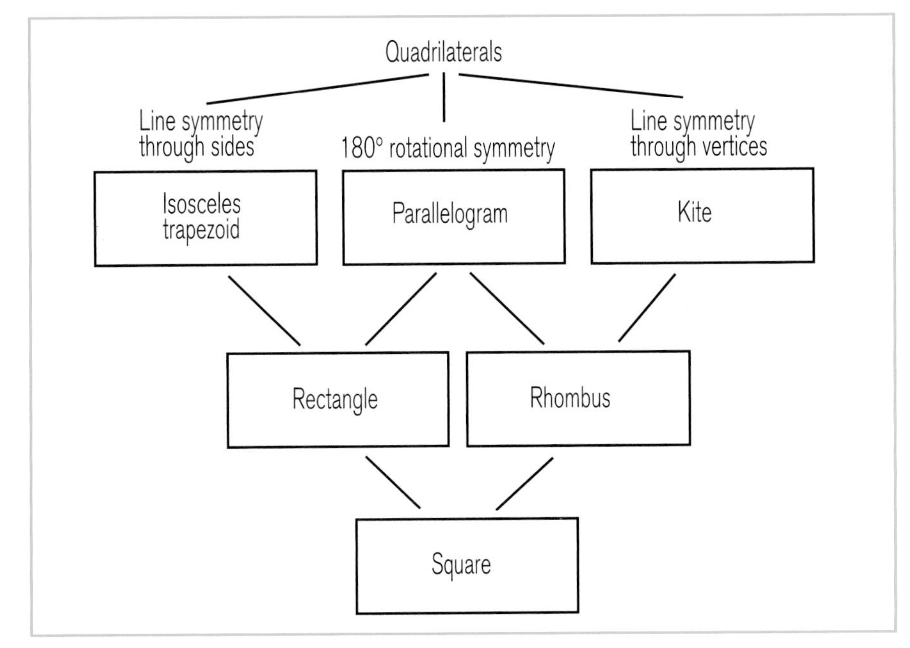

Fig. **3.5.**

A classification scheme for quadrilaterals that is based on types of symmetry

Students in the middle grades often have difficulty understanding why a quadrilateral can have more than one name. They find it confusing when asked to use *all* applicable names for geometric shapes. This confusion makes hierarchical classification a difficult topic. Students tend to identify a shape by its most obvious features rather than analyze it to determine if it also displays the characteristics of a broader category of shapes. For example, the fact that a square has two lines of symmetry through its vertices is sufficient to call it a rhombus. However, the mental picture most students have of a rhombus as having no right angles overshadows this characteristic. Using symmetry as a test for identifying a rhombus can help students make their notion of the category more inclusive.

Conclusion

The exemplars in this chapter illustrate important principles of transformations and symmetry. The varied experiences are essential to provide students with a broad range of opportunities to develop their spatial-visualization skills and their ability to reason geometrically.

Students should develop an understanding of how translations, rotations, and reflections work. The activities in the chapter address several goals for middle-grades geometry recommended in *Principles and Standards for School Mathematics* (NCTM 2000): "describe sizes, positions, and orientations of shapes under informal transformations ... [and] examine the congruence, similarity, and line or rotational symmetry of objects using transformations." These transformation-geometry activities give students another lens for investigating and interpreting geometric objects.

Chapter 4

Visualization, Spatial Reasoning, and Geometric Modeling

Important Mathematical Ideas

"Students' skills in visualizing and reasoning about spatial relationships are fundamental in geometry."
(NCTM 2000, p. 237)

Our daily activities require us to deal constantly with spatial orientation. Visualization and reasoning about spatial relationships are necessary to our everyday functioning. Research has found a strong correlation between spatial ability and problem-solving performance, suggesting that spatial visualization is a good predictor of successful problem solving (Tillotson 1984; Geddes and Fortunato 1993).

Principles and Standards for School Mathematics (National Council of Teachers of Mathematics [NCTM] 2000) identifies the following as important abilities for students in grades 6–8. They should—

- draw geometric objects with specified properties, such as side lengths or angle measures;
- use two-dimensional representations of three-dimensional objects to visualize and solve problems such as those involving surface area and volume;
- use visual tools such as networks to represent and solve problems;
- use geometric models to represent and explain numerical and algebraic relationships;
- recognize and apply geometric ideas and relationships in areas outside the mathematics classroom, such as art, science, and everyday life. (P. 232)

Many middle-grades students enjoy reading Flatland *(Abbott 1937), a novel that raises questions about space and how we see things. Literature such as this novel can stimulate students' interest in geometry.*

What Might Students Already Know about These Ideas?

Students in the lower elementary grades have experiences working with physical representations of shapes. As they describe and change the attributes of shapes, they begin to develop important spatial-visualization and reasoning skills. Specifying landmarks and identifying structures from certain perspectives builds on this foundation. In the upper elementary grades, students' work with two- and three-dimensional shapes helps them learn the characteristics of those shapes. They become skilled in various ways to represent such shapes. They build models and use many manipulatives to explore both numerical and geometric relationships. Exploring relationships and testing conjectures should be mainstays of their mathematics experiences.

You can informally assess your students' prior experience in spatial reasoning and visualization with activities using tangrams. The following exercise is designed to aid in this assessment.

Exploring Shapes with Tangrams

Goals

- Assess students' spatial-reasoning and visualization skills
- Reinforce and develop spatial-visualization skills
- Explore the properties of figures created with tangram pieces

Materials and Equipment

- A set of tangrams for each student
- Paper for drawing figures

A template for tangrams can be found on the CD-ROM. Heavy paper stock is well suited for reproducing the tangram pieces.

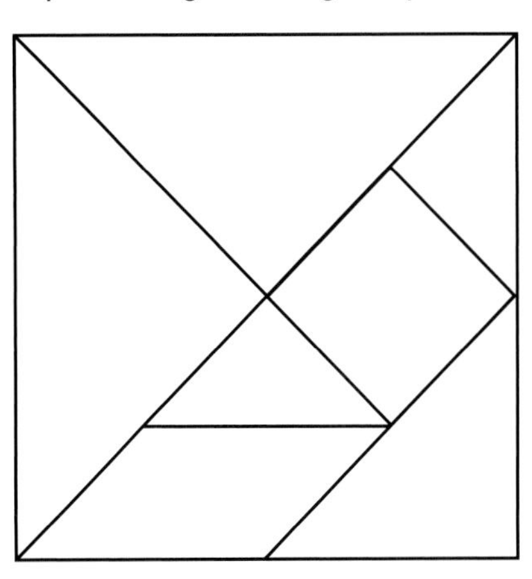

Activity

Give the students any or all of the following tasks and ask them to draw the resulting figures on paper:

1. Use any number of tangram pieces to form as many different squares as possible.
2. Use tangrams to form as many different composite triangles as possible.
3. Explain which of the tangram triangles (individual triangles or composite triangles) are similar.
4. Make the designs in figure 4.1 using all seven tangram pieces.

Fig. **4.1.**

Designs created with tangrams

The interactive explorations Tangram Puzzles and Tangram Challenges are available on NCTM's Web site (address: nctm.org; click on "online version" of *Principles and Standards* and then on "Electronic Examples," and scroll down to example 4.4.)

Discussion

This activity gives students an opportunity to demonstrate flexibility in manipulating shapes as they slide, turn, and flip the various tangram

"The Tangram Conundrum" (Thatcher 2001) offers a rich discussion of using tangram pieces to reinforce mathematical concepts taught in the middle grades. It includes a discussion of why a square cannot be created from six tangram pieces.

pieces. They must be aware of the relationships among different tangram pieces to match edges and form angles. As the students discuss their work, they may mention congruent sides, complementary and supplementary angles, motions, and orientations. In addition, the students' drawings of their composite shapes can give the teacher information about their ability to use diagrams to represent their thinking.

As with other topics, it is important in spatial visualization to consider students' levels of geometric thinking. Some students at the middle-grades level can identify a shape and reproduce it, but they may experience difficulty putting together shapes to form other shapes. Most students can see relationships among component parts of a figure and experience success in making composite figures with the various pieces of the tangram puzzle. They may use appropriate language when putting together sides that have the same length and when describing the angles of various pieces and how they fit together. They may use terms such as *congruent* and *similar* correctly. Students who have well-developed geometric skills will be able to make arguments about similarity in the various tangram shapes.

Developing Spatial Visualization

Creating tessellations is a challenging and exciting activity for middle-grades students. The interactive applet Tessellate! from Project Interactive (www.shodor.org/interactivate/activities/tessellate/) explores the tessellation of figures in the plane, which can help students develop their spatial-visualization skills.

This section includes several experiences that will help middle-grades students sharpen their spatial-visualization skills. Visualization is an important consideration for commercial advertising. Most company logos display some form of symmetry to ensure that the logo is pleasing to potential clients. Companies incorporate geometric shapes as well as such geometric concepts as rotations, symmetry, and dilations in the design of their logos. Ask the students to brainstorm about the various geometric properties they see in the logos with which they are familiar. The various geometric characteristics make certain logos easy to recall and readily identifiable with the corporation. The logo for Mercedes-Benz, for example, shows a number of geometric properties, including line symmetry and rotational symmetry. It employs a circumscribed circle and displays congruent radii. The following exercise, Logos and Geometric Properties, requires students to identify the specific geometric properties in corporate logos.

Identify the types of symmetry found in the Mercedes-Benz logo, shown at the right. What other geometric properties does it have? Logos are excellent examples of how businesses use geometry, particularly in advertising.

Courtesy of Daimler-Chrysler AG and its registered subsidiary Mercedes-Benz USA, One Mercedes Drive, P.O. Box 350, Montvale, NJ 07645-0350.

Logos and Geometric Properties

Goals

- Develop spatial-visualization skills
- Explore geometric relationships used in business and advertising
- Design figures with geometric properties

Materials and Equipment

- A copy of the blackline master "Logos and Geometric Properties" for each student
- Logos that students have clipped from magazines, newspapers, or commercial products and brought to class from home

p. 112

Activity

The students divide their logos into categories according to the symmetries or other properties they display or the shapes they include. They then create their own logo that displays at least two geometric properties, identify the properties, and describe how the logo demonstrates each one.

Discussion

By identifying logos that display various geometric figures and concepts, the students apply those concepts in a different setting—advertising media. The students could also select a logo and then create a larger- or smaller-scale version of it. You could display the students' creations, along with their cutout logos, by category on the wall of your classroom or on a bulletin board.

Drawing Geometric Objects

Students need multiple and varied experiences in drawing and interpreting two- and three-dimensional geometric objects that demonstrate certain properties. Activities that involve representations such as floor plans or scale models require students to think about such properties of figures as lengths of sides and angle measures, and they afford students an opportunity to apply geometry outside the classroom (NCTM 2000). In the exercise Picturing Ziggurats (based on spatial-visualization materials in *Ruins of Montarek* [Lappan et al. 1996]), students consider a base plan for a structure made from blocks. Interpreting base plans by building the depicted structures with blocks and by drawing front, side, or back views of the structures requires students to consider the properties of various geometric figures to verify the information. Tasks such as this one develop an appreciation and understanding of geometric properties.

Isometric drawings are another common way to model geometric structures. Isometric dot paper is a grid system that is useful in drawing

A blackline master for isometric dot paper is available on the CD-ROM that accompanies this book.

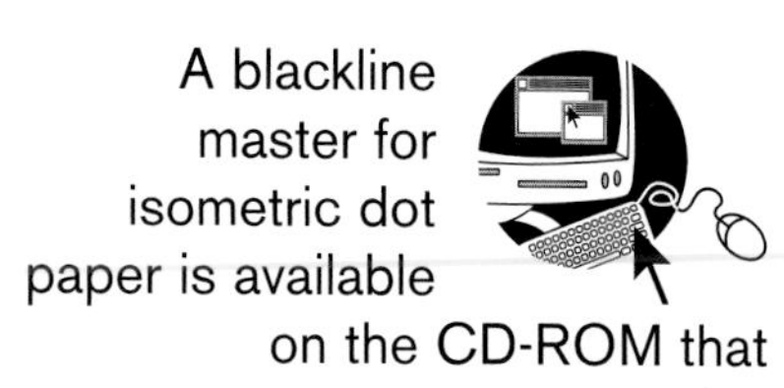

Fig. **4.2.**

An illustration of the six ways one dot can be connected to an adjacent dot on isometric-dot paper

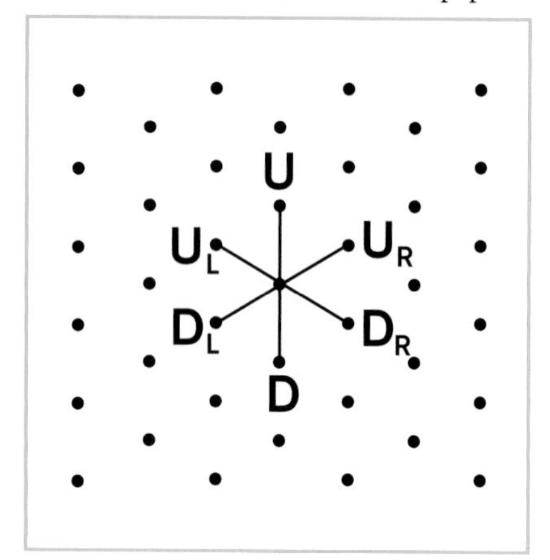

two-dimensional representations of three-dimensional figures, which helps students develop spatial-visualization skills. The dots on isometric paper are arranged on a triangular, rather than a square, grid, with equal distance between adjacent dots. As depicted in figure 4.2, six "moves" are possible that connect the vertices of objects drawn on isometric dot paper. Identifying the following moves may be helpful to students (Winter, Lappan, Phillips, and Fitzgerald 1986, p. 67):

U: Drawing up
D: Drawing down
U_R: Drawing a line up and to the right
U_L: Drawing a line up and to the left
D_R: Drawing a line down and to the right
D_L: Drawing a line down and to the left

Picturing Ziggurats

Ruins of Montarek invites students to engage in extensive explorations with isometric coordinate planes. Among other things, they learn how to represent a structure on grid paper by reading a set of "building plans." These building plans consist of a base map of a block structure and information about how many blocks are stacked in a particular spot. Special arrangements of stacks of blocks are called *ziggurats.* For example, figure 4.3a shows a base plan (the "footprint") of a row of adjacent buildings ranging from four stories high to one story high. In this example, there are four stacks of blocks, or "stories." The number in each cell in this footprint indicates the number of blocks stacked above that cell. Figure 4.3b shows what this arrangement of buildings would look like from the front view. As an introduction to the following activity, Isometric Explorations (used with permission from *Ruins of Montarek* [Lappan et al. 1996], pp. 87–88), the students might benefit from creating this structure using cubes or other building materials.

Isometric Drawings

As a follow-up to the activity with ziggurats, students should draw representations using isometric dot paper. To begin, the teacher might construct a model and ask the students to draw the isometric representation. Once the students have some experiences with creating such drawings, they could work in pairs to build structures with cubes or other materials and to draw two-dimensional representations of them on isometric dot paper.

Fig. **4.3.**

(a) The base map, or "footprint," of a row of buildings and (b) the front view of the structure depicted in the footprint

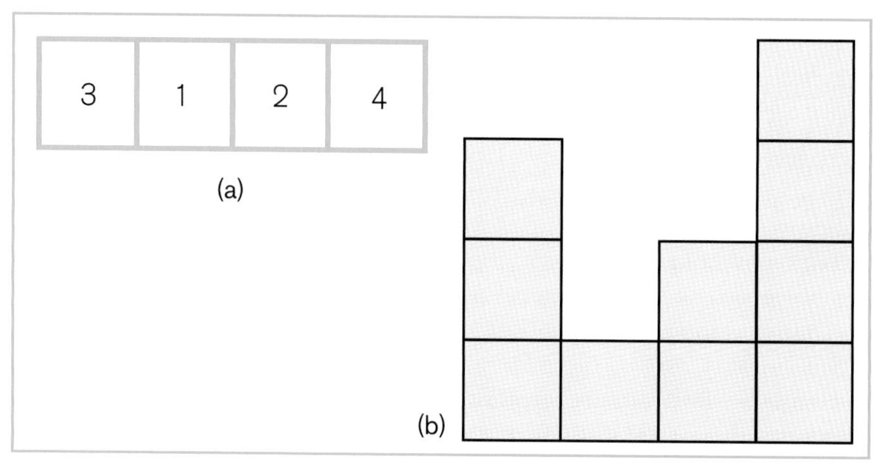

Isometric Explorations

Goals

- Represent structures in isometric drawings
- Use spatial visualization to identify isometric representations of base plans
- Use spatial visualization to identify various two-dimensional views of isometric drawings

Materials and Equipment

- A copy of the blackline master "Isometric Explorations" for each student
- Rulers
- Cubes or other construction materials

p. 113

Activity

In this activity, students match base maps with various isometric views of structures and isometric drawings with various two-dimensional representations of building views. They also draw a specified isometric view of a structure from a base map of the structure.

Discussion

Cubes or other construction materials should be available for students who need physical materials to complete this investigation. You may want the students to work in pairs to discuss their responses or use cubes to construct the buildings to help them determine the correct perspectives. For the final question, you may wish to have students in different groups work on different perspectives and then have them draw other views and share their drawings. Students strengthen their spatial reasoning by discussing and justifying their approaches.

Cross Sections of Solids

One activity that involves students in analyzing the properties and characteristics of two- and three-dimensional shapes is determining the cross section that is formed when a plane intersects a solid. This task can be difficult for many students because it involves describing or drawing a two-dimensional figure that results from intersecting a three-dimensional figure with a plane. Such experiences with three-dimensional figures, however, can help students develop their ability to visualize solids. You might ask your students, for example, what shapes can be formed by different cross sections of a cube. Students can readily see that a square is one cross section of a cube. What about a rectangle, a triangle, a trapezoid, a pentagon, or a hexagon?

Students can begin the following activity, Cross Sections of Three-Dimensional Shapes, by exploring the cross section formed when a plane intersects a cube to form a square. Discuss or demonstrate this

Fig. **4.4.**

A square cross section of a cube

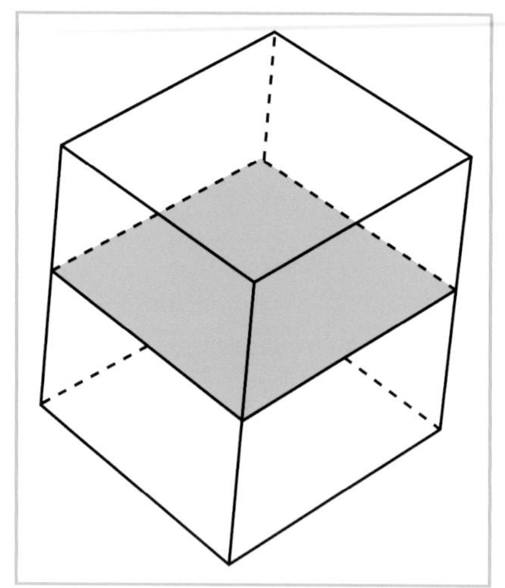

result before the students proceed with the rest of the activity. The students can benefit from describing the characteristics of the cross sections and their relationship to the characteristics of the cube. Ask them to draw the resulting cross section, both as a single geometric figure and as part of the original solid (see fig. 4.4).

Cross Sections of Three-Dimensional Shapes

Goals

- Explore cross sections of solids
- Develop spatial-visualization skills

Materials and Equipment

- A copy of the blackline master "Cross Sections of Three-Dimensional Shapes" for each student
- Rulers and paper for drawing
- Models of solids (if available)

p. 115

Activity

Students draw the cross sections formed when various solids are intersected by a given plane.

The applet Spinning and Slicing Polyhedra on the CD-ROM can be used in conjunction with this activity.

Discussion

The students may have difficulty visualizing the results of the intersections. You may select several of these exercises for a warm-up and then continue with the others. Having students model, describe, and draw one or more of the cross sections will help them be more successful with the more difficult problems. The importance of physical models when studying cross sections cannot be overemphasized. Students who have had little experience with spatial-visualization activities will likely have difficulty visualizing these cross sections without a model because they cannot envision the unseen faces of the shapes (NCTM 2000, p. 237). Each of the figures could be constructed from clay. You might talk with the art teacher about joint projects or sharing resources. As an extension, the students could make a model of a solid and demonstrate one or more cross sections that are not included on the activity sheet. The students could work alone or in groups and later share their products with the rest of the class.

Using Two- and Three-Dimensional Representations

Spatial visualization is the ability to understand and perform imagined movements on objects in two- and three-dimensional space (Clements and Battista 1992). Spatial orientation is the ability to operate in space and understand one's own position in space with respect to the location of other objects. This ability is exhibited in, for instance, finding one's way. Tasks that involve spatial visualization and orientation can be confusing for students when they require translating between a three-dimensional object and a two-dimensional representation—for instance, when interpreting a diagram or making a drawing of three-dimensional structures or locations in the real world. For example, middle-grades students often have trouble representing on isometric dot paper a building they have constructed with centimeter cubes. They struggle to transform (e.g., rotate) an object mentally, to imagine and draw it from different perspectives, or to inspect it mentally in order to answer questions about it. To help students overcome such difficulties, you can offer them exercises that involve both spatial visualization and spatial orientation. The next activity, I Took a Trip on a Train, which appears on the CD-ROM (courtesy of the Annenberg/CPB Project), is one such exercise.

I Took a Trip on a Train

Goals

- Understand the attributes and component parts of two-dimensional representations of three-dimensional objects
- Develop spatial-visualization skills by analyzing perspective views of three-dimensional objects
- Use spatial reasoning to position objects in a perspective view

Materials and Equipment

- Computers at which individual students or small groups of students can work
- I Took a Trip on a Train (activity on the CD-ROM that accompanies this book)

Activity

The activity involves mentally picturing different views of the same arrangement in space. The students must connect a two-dimensional bird's-eye view of a set of objects, such as might be represented in a map, to two-dimensional representations ("photographs") of the three-dimensional views one would have from the ground. The students are presented with a map of a region with several geometric objects indicated. They imagine that they have taken a trip on a circular train track and have taken a series of four photographs while looking toward the center of the region. The photos are not in the order in which they were shot, so the students' task is to order the photographs.

See "Cube Challenge," by Judy Bippert, on the CD-ROM. The article includes extended spatial-visualization investigations involving cubes.

Discussion

Students who are having trouble ordering the photographs can sketch an enlarged map on chart paper, place it on a table or the floor, and physically walk around it. Questions such as "If you were here [*indicate a particular spot on the circle*], would the low shed appear to be to the right or to the left of the spike?" will help the students compare what they imagine with a particular photo.

Extending Students' Explorations with Two- and Three-Dimensional Figures

In the following activity, Constructing Three-Dimensional Figures, students could use manipulatives such as Polydrons to build and draw polyhedra and to draw their nets. Manipulatives such as Geofix shapes and Polydrons are recommended for activities that involve building figures and exploring spatial relationships. Alternatively, the students could use coffee stirrers, toothpicks, or straws to form edges and gumdrops, marshmallows, or raisins for the vertices. The resulting figures

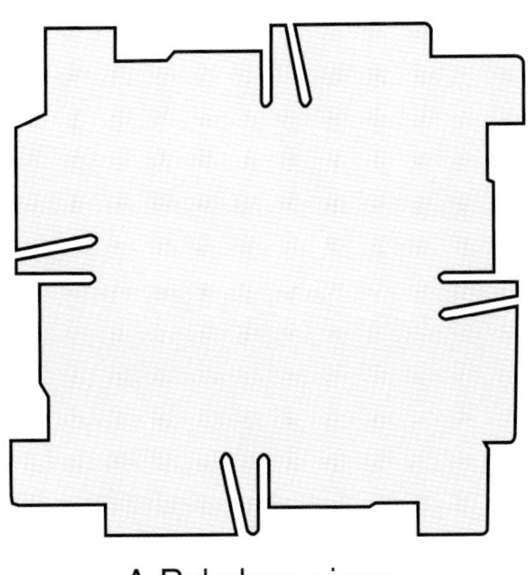

A Polydron piece

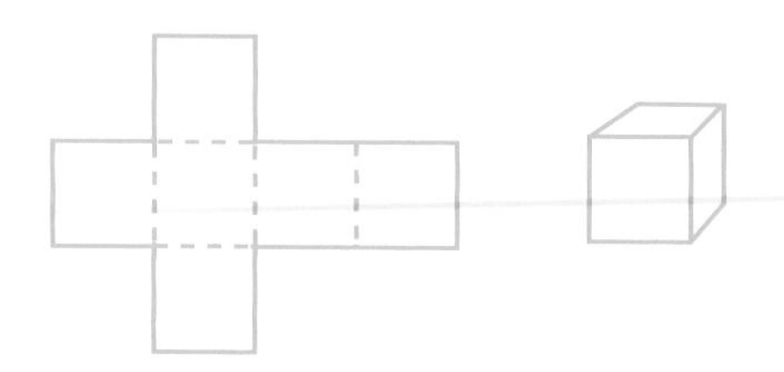

A *net* is a two-dimensional representation used to create a three-dimensional shape. The net can be folded into the three-dimensional object.

are not as precise as those built with commercial products, but they do enable the students to study the properties of solids.

Although the curriculum in grades 3–5 calls for students to have experience with three-dimensional shapes, this activity may be some students' first encounter with building such shapes. Concrete materials give students the opportunity to touch and feel the faces, edges, and vertices of various polyhedra. They can turn the polyhedra in their hands and view them from different perspectives. Such hands-on experience helps students make more-accurate drawings, on either plain or isometric dot paper. They can disassemble a polyhedron to form a net and then try to find other nets of the same shape.

Constructing Three-Dimensional Figures

"Students also need to examine, build, compose, and decompose complex two- and three-dimensional objects."
(NCTM 2000, p. 237)

Goals

- Investigate characteristics of solids
- Develop spatial-visualization skills through constructing and disassembling shapes

Materials and Equipment

- A copy of the blackline master "Constructing Three-Dimensional Figures" for each student or group of students
- Materials suitable for building solids that can be opened to form nets, such as Geofix shapes and Polydrons
- Paper for drawings
- Rulers

p. 116

Activity

With Geofix shapes, Polydrons, or other similar materials, the students build two given solids, record information about some of the properties of the solids, draw nets of the solids, and describe the relationship between the net and the solid. They also build and sketch solids using only specified shapes. The students may work individually or in groups to complete the tasks. Remind them that they should be prepared to explain their methods and answers.

Discussion

Constructing and manipulating three-dimensional shapes and interpreting two- and three-dimensional representations extend students' experiences with shapes. The sequence gives them an introduction to three-dimensional shapes and the experience necessary to begin to analyze them.

The students can explore various geometric properties with the solids they have constructed. Ask them to find and describe examples of the following concepts: lines of symmetry for a face, two parallel lines determining a plane, two intersecting lines determining a plane, three noncollinear points determining a plane, two planes intersecting in a line, and noncoplanar points.

Refer to "Exploring Relationships with Polyhedra" in chapter 1. In that task, the students collect data about the number of faces, vertices, and edges of various polyhedra and make a conjecture about the relationship among the numbers (Euler's formula). If the students have not explored that relationship, consider using that investigation as a prelude to this activity.

For additional reading and ideas, see "Visualizing Three Dimensions by Constructing Polyhedra," by Victoria Pohl. This paper, from the NCTM 1987 Yearbook, Learning and Teaching Geometry, K–12, *is included on the CD-ROM.*

Using Visual Tools to Represent and Solve Problems

Geometric models can play an important role in students' mathematics development because of their usefulness in illustrating specific relationships. They can also help illuminate and clarify concepts and serve as a framework to solve problems in real-world settings. Geometric models may be as concrete as any of a variety of familiar manipulatives or more abstract, like the geometric models for understanding probability relationships.

Geometric models are useful in helping students understand the relationship between area and perimeter. In lower grades, it is not uncommon for students to confuse the two terms and concepts. But by sixth grade, students should have a clear understanding of the two concepts. However, many middle-grades students erroneously think that they are directly related; that is, they think that when two figures have the same perimeter, they have the same area or that if the areas of two figures are the same, then their perimeters are the same. Although area and perimeter are distinct concepts, there is a relationship between them that the models (a table and a graph) in the next activity reveal.

Minimizing Perimeter

Goals

- Explore the relationship between perimeter and area
- Use geometry for real-world applications

Materials and Equipment

- Grid paper (available on the CD-ROM)
- Centimeter cubes

Activity

This activity might be done in pairs or small groups. First, ask the students to find the least amount of fencing for a rectangular garden plot that is 36 square feet in area. Have them organize the information in a table like the one in figure 4.5. Second, ask the students to graph perimeter vs. length for their data. Next, have them repeat this process with an area of 24 square feet. For simplicity, tell them to use only whole-unit dimensions. Finally, ask them to make a conjecture about the minimum fencing needed for an area of 100 square feet and to write a paragraph defending their conjecture.

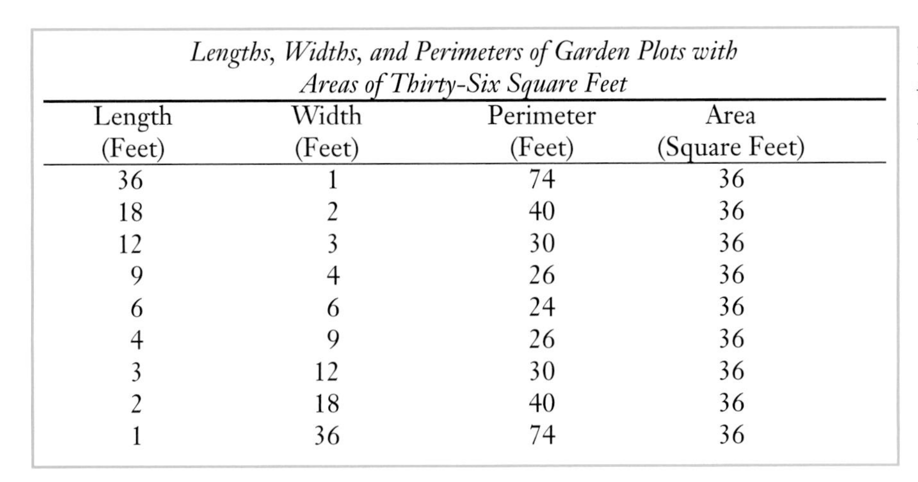

Lengths, Widths, and Perimeters of Garden Plots with Areas of Thirty-Six Square Feet

Length (Feet)	Width (Feet)	Perimeter (Feet)	Area (Square Feet)
36	1	74	36
18	2	40	36
12	3	30	36
9	4	26	36
6	6	24	36
4	9	26	36
3	12	30	36
2	18	40	36
1	36	74	36

Fig. **4.5.**
Values for the problem about minimizing perimeter

Discussion

As the table and the graph in figure 4.6 show, the 36-square-foot rectangular garden plot requiring the least amount of fencing is a square with side length 6 feet. The students' investigations with an area of 24 square feet should determine that the whole-unit dimensions resulting in the smallest perimeter are 4 feet by 6 feet, so the least perimeter is 20 feet. For a 100-square-foot area, the minimum perimeter is 40 feet, and the dimensions are 10 feet by 10 feet. In their conjectures and defenses, the students should express an understanding that area does not necessarily increase as perimeter increases. They should also conclude that for a fixed rectangular area, the perimeter is minimum when the shape is a square.

For the area of 24 square feet, students can also explore fractional values for length and width to discover that the area is a minimum when the shape is a square with sides of $\sqrt{24}$ feet (about 4.9 feet). The minimum perimeter, then, is about 19.6 feet.

Extensions

As an extension, the students could explore the relationship between the surface areas and the volumes of similar solids. One set of

Students could also investigate how the area varies if the perimeter is held constant.

Fig. **4.6.**

A graph of the rectangular-garden-plot data

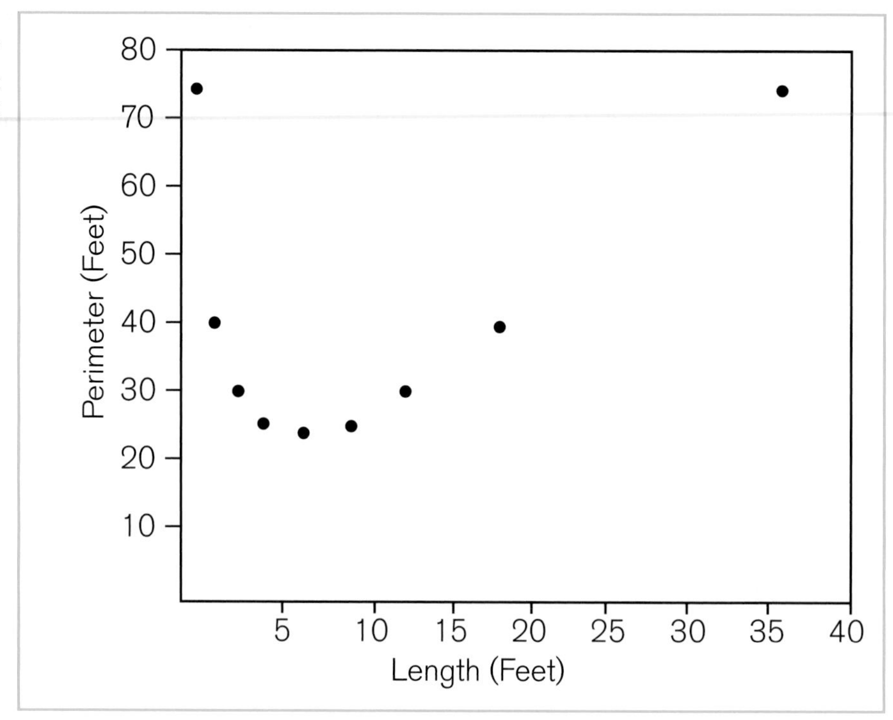

relationships that geometric models can help students understand is the relationship among the lengths of the sides, the surface areas, and the volumes of similar solids. Proportional reasoning can be a challenge for students, and proportional relationships between corresponding sides of similar figures can prove difficult for them to understand. The ratio of the lengths of any of the corresponding edges of two similar solids is the same as the ratio of the lengths of any other pair of corresponding edges. This ratio is not the same as the ratio of their surface areas or of their volumes. If the ratio of the lengths of any two corresponding edges of two similar solids is $n:m$, then the ratio of their surface areas is $n^2:m^2$ and the ratio of their volumes is $n^3:m^3$. The quadratic relationships between edge lengths and surface areas of similar solids and the cubic relationships between edge lengths and volumes of similar solids can be counterintuitive and thus difficult for students to understand. The computation of such nonlinear relationships can add another layer of difficulty. To help the students understand these relationships, have them create a table showing the surface areas and volumes of cubes with various edge lengths, as illustrated in the table in figure 4.7. Once the students have completed the table, they can begin to examine the relationships among side lengths, surface areas, and volumes.

Fig. **4.7.**

A table that might help students discover the relationships among the lengths of the edges, the surface areas, and the volumes of cubes

Edge Lengths, Surface Areas, and Volumes of Cubes

Length of Edge (Units)	Surface Area (Square Units)	Volume (Cubic Units)
1	6	1
2	24	8
3	54	27
4	96	64
5	150	125
6	216	216
8	384	512
10	600	1000

The following questions can help organize the discussion:

Surface Area

1. When the length of an edge of a cube doubles, how does the surface area change? (It increases by a factor of 4.)
2. What happens to the surface area of a cube when an edge becomes three times longer? (It increases by a factor of 9.) Four times longer? (It increases by a factor of 16.) Half as long? (It becomes 1/4 the original.)
3. By what factor does the surface area of a cube increase when an edge is increased by a factor of 10? (100) By a factor of 20? (400) By a factor of n? (n^2)

Volume

1. When the length of an edge of a cube doubles, how does the volume change? (It increases by a factor of 8.)
2. What happens to the volume of a cube when an edge becomes three times as long? (It becomes twenty-seven times as large.) Four times as long? (It becomes sixty-four times as large.) Half as long? (It becomes 1/8 the original.)
3. By what factor does the volume of a cube increase when an edge is increased by a factor of 10? (1000) By a factor of 20? (8000) By a factor of n? (n^3)

Students should be encouraged to build centimeter-cube models of various cubes in the table and record the number of centimeter cubes required, using an edge, a face, and the volume of a centimeter cube as a unit, a square unit, and a cubic unit, respectively. By building models to examine the changes in surface area and volume as the lengths of the edges of the cubes change, students can begin to gain an intuitive sense of these nonlinear relationships. The actual relationships—quadratic and cubic—can be better understood and remembered by using models to represent the cubes.

Once the students have completed the table and can express the relationships among the measures of the edges, surface areas, and volumes of similar solids, they might describe how doubling the size of an aquarium affects the amount of water and the number of fish it will hold, the weight, and so on. An aquarium with edge lengths that are doubled will require eight times as much water and will consequently weigh eight times as much as the original aquarium. It will accommodate eight times the number of fish.

Spatial Reasoning with Circles

The following problem, from Silver (2000, p. 21), available on the CD-ROM, connects to topics in algebra, geometry, proportionality, computation, and measurement:

> Two students are considering the diagrams shown in the margin on the next page, in which circles are inscribed in a square with a side length of 8 inches. Carlos says that the shaded portion in diagram C is larger than the shaded portion in diagram B or A. Bill disagrees, asserting instead that the shaded

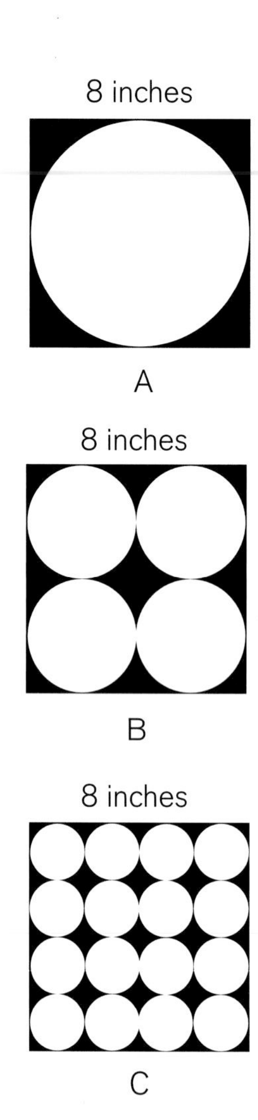

portion in diagram A is larger than the shaded portion in diagram B or C. Alicia disagrees with both boys. She says that all three diagrams have the same portion shaded. Who is correct? Give a mathematical justification for your answer.

By giving a mathematical justification for this problem, students generalize by writing an algebraic formula that applies to a square of any size and to any number of equal subdivisions. The exercise also gives students experience with argumentation by having them present a mathematical justification or informal proof that the total area of the shaded region remains the same regardless of the number of times the square is subdivided according to the pattern illustrated. The problem also presents an example of proportional relationships. The ratio of the areas of similar plane figures is—for polygons—the square of the ratio of the lengths of their corresponding sides or—for circles—the square of the ratio of the lengths of their radii. Experiences with proportionality in such contexts assist students in understanding and recognizing proportional relationships.

Another justification of the outcome of this problem uses dilations. If the students have performed dilations and can identify and understand their properties, they can show that the areas of the shaded regions in all three figures are the same because the circles and the squares within which they are inscribed are dilations of one another. This problem exemplifies how middle-grades students can use empirical experience to develop arguments.

The study of geometry offers rich experiences with connections not only to other areas of mathematics but also to various aspects of life both in school and outside school. Such connections help students appreciate the importance of geometry in the world.

In the next activity, Indirect Measurement, students learn to apply the principles of similar triangles to solve problems by taking indirect measurements.

Indirect Measurement

Goals

- Apply similarity to problems
- Use indirect measurement

Materials

- A copy of the blackline master "Indirect Measurement" for each student
- Rulers

Activity

A method of using similar triangles to find the distance across a rock formation is described, and two problems involving finding the width of natural landscape features are presented. The students are then asked to describe a method of finding the height of a tall tree if the height of a smaller one is known.

Discussion

Both the rock-formation problem and the pond problem deal with similar triangles. The students should verify that the markers for line segments *DE* in the rock-formation problem and *BC* in the pond problem were placed so that the corresponding base angles of the smaller and larger triangles are congruent, which results in two similar triangles. The students should note that similarity of the triangles means that $\overline{BC}$ and $\overline{DE}$ are parallel. They should also be able to name the corresponding sides of the similar triangles. Have them find the missing measures and identify the scale factor for each figure. The students might use the ratio 80/100 in the rock-formation problem to determine a scale factor of 4/5. They might also set up the proportion $100/60 = 80/x$ to find the length of $\overline{BC}$ to be 48 feet. Doing so gives 48/60, or 4/5, as the ratio of the corresponding sides. In the third problem, prompt the students by asking, "What would you need in order to find the height of the trees?" Suggest that they consider the shadows cast by the trees. For example, assume that the shadow cast by the larger tree is twenty-five feet and the shadow cast by the smaller tree is five feet.

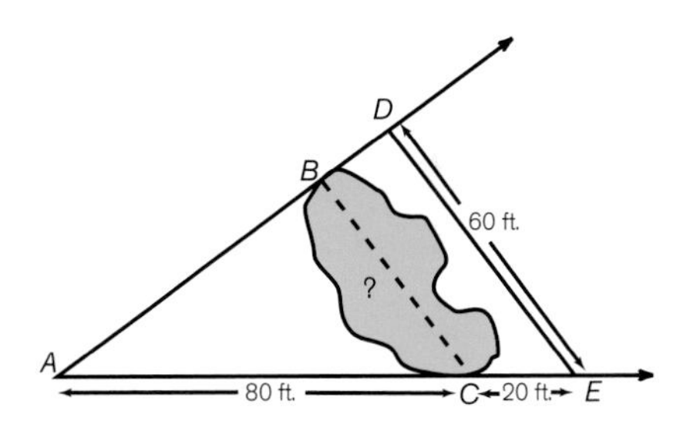

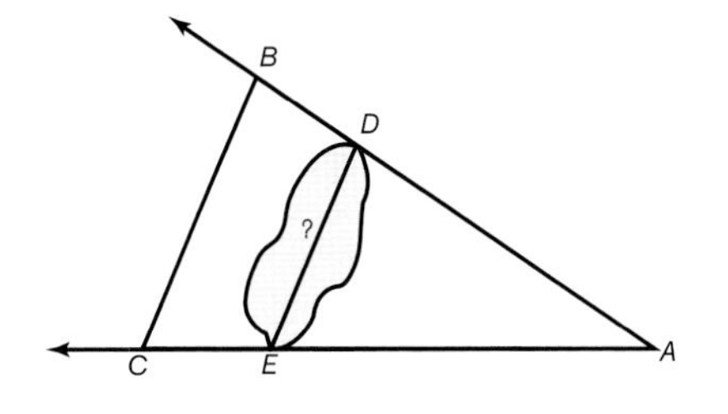

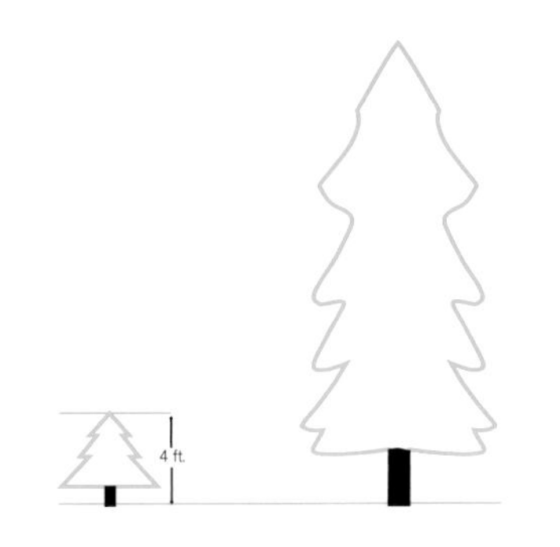

Problems such as those dealing with similar triangles demonstrate the importance of geometry in problem solving. Students of all ages should expand their understanding of the importance of geometry in their lives, so problems presented to students should emphasize the importance of geometric principles and reasoning. The following activity, The Race, engages students in solving a problem that depends on applying geometric principles.

The Race

Goals

- Explore geometric relationships
- Use geometry to solve problems

Materials and Equipment

p. 119

- A transparency copy of the blackline master "The Race" or a copy of the race diagram drawn on the board
- An overhead projector

Activity

Display the transparency of "The Race" on the overhead projector. Tell the students that a problem emerged when several students in Mr. Nadtando's class wanted to determine who was the fastest runner. They had a large square already outlined on their playground, so they decided that they would use it as the location for the race. They agreed on some conditions:

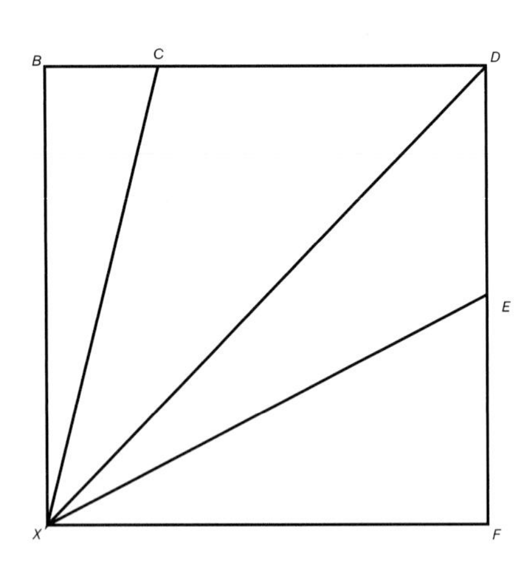

- All runners would start at the same place (marked by *X*).
- The following were the assignments for the course: Blanca would run to point *B;* Candy would run to point *C;* Demetrius would run to point *D;* Eli would run to point *E;* and Fran would run to point *F*.
- The first person to reach the spot indicated by his or her letter would be the winner.

Explain that the projected diagram shows the way the students set up the field. Continue to recount that as the students were about to begin the race, Demetrius stopped and said that he believed the race would be unfair because he would have to run farther than the others. Ask the students, "Is Demetrius right? How would you justify agreeing or disagreeing with Demetrius? If you were to design the race so that it would be fair to everyone and maintain the original conditions, how would you set up the field? Make sure you explain your proposed plan so that your classmates will be able to understand that it is fair."

Discussion

In general, students quickly recognize that the original arrangement is unfair. They can visualize that the distances are not equal. Most students talk about "measuring the lines" to show that the lengths are not equal. Some students show an initial understanding of geometric principles that support their conclusions. The following responses came from a seventh-grade class:

- Yes, I agree because he is going diagonal.
- Yes, because he has to run the hypotenuse and the others only have to run a shorter length.
- They're all triangles, but Demetrius's triangle has the longest sides, and he is running the hypotenuse, which is always longer.

Encourage the students to present their alternative plans and to justify them. The diagrams in figure 4.8 are the plans that several students constructed to respond to the problem. The majority of the students used the first approach (fig. 4.8a), which solved the problem of unequal distances but did not keep a common starting point. The plan also makes an assumption that the lines are parallel. Can the students who propose such a plan explain how they can be sure that the distances are equal? The other three diagrams maintain a common starting point. The second plan (fig. 4.8b) does not remedy the unequal-distances problem. All the runners except the one going to *A* run equal distances only if *X* is in the center. Can the students explain how they know the point is in the middle by, for instance, using diagonals? Do they see that the runner to *A* has the same problem that concerned Demetrius? The last two diagrams meet all requirements for a fair race by using circles. What are the benefits of each type of course? What geometric justification can be used to argue that the distances are all equal (such as that the radii of a circle are congruent)? It is important to provide students with an opportunity to justify their plans and to assist them in using geometric properties and principles to support their arguments.

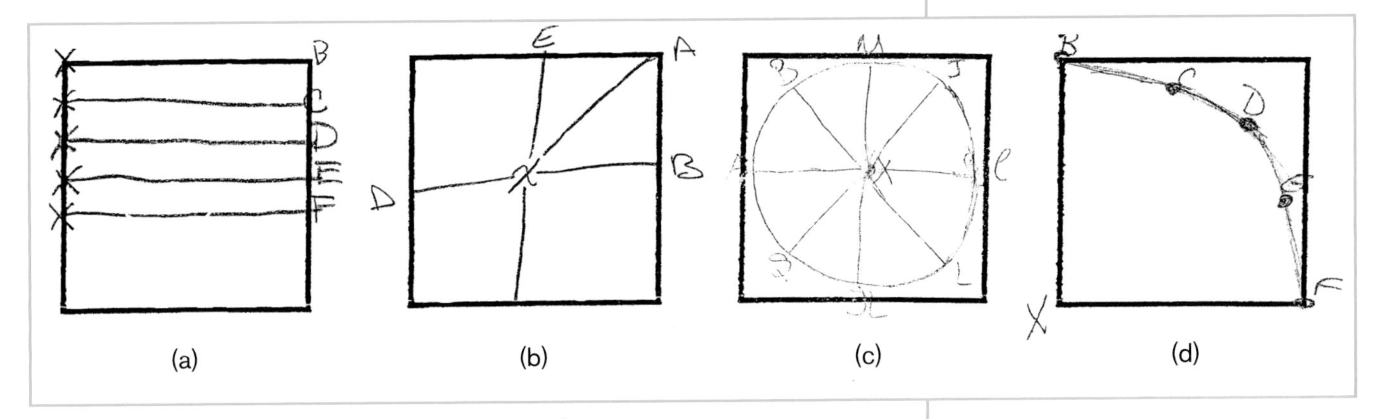

Fig. **4.8.**

Several diagrams showing students' soltions to the race problem

Conclusion

The activities and examples in this chapter were selected to offer students experiences that are important in extending their spatial-visualization skills. It is important in implementing such activities that students develop an understanding of the relationships between two- and three-dimensional figures and that they apply geometric properties and principles in solving problems. The activities address important goals from *Principles and Standards for School Mathematics* (NCTM 2000): drawing and representing two- and three-dimensional figures, understanding the relationships between two- and three-dimensional figures, and using visualization skills to represent and solve problems. As emphasized throughout this book, it is important for students to develop sound reasoning as they represent geometric ideas and apply geometry to problem situations.

GRADES 6–8

NAVIGATING *through* GEOMETRY

Looking Back and Looking Ahead

The middle grades are important in students' geometric development and are necessary in helping students become observers of the world around them who explore the geometric relationships, measures, and motions in their environment.

Students bring a complex understanding of basic geometry and spatial relationships to school at an early age.

Geometry in the middle grades is essentially a bridge. It transports students from the early development of geometric reasoning and spatial relationships in the elementary grades to the logical, reasoned proofs and study of three-dimensional objects in coordinate space at the secondary level. The activities throughout this book provide tools for moving students ahead in their geometric thinking. You are encouraged to continue thinking about the issues raised in this book and the questions you may have formulated as your students solved the various problems in the activities.

When students enter kindergarten, they bring with them a wide range of knowledge. They can talk and, perhaps, read. They likely have memorized the numbers up to ten and the letters of the alphabet. They also bring with them an understanding of geometry that is not easily discernible. Although primitive, their geometric understanding allows primary school students to position themselves in their familiar world. Despite the fact that they may lack the vocabulary to express it, young children can sense their position and size relative to their bed, the kitchen table, their school desk, and their teacher.

In elementary school, these primitive but emerging understandings are clarified, expanded, and deepened. Through a variety of explorations with manipulatives and dynamic geometry software, students can begin to recognize and analyze geometric figures, as well as compose and decompose them (NCTM 2000, p. 237). In addition, students in the lower grades develop the ability to recognize and describe a variety of spatial relationships (e.g., location, direction, and size), and they develop such skills as representation, identification, and categorization of figures. It is the task of the middle-grades teacher to build on these skills and understandings.

During the middle grades, students develop the ability to analyze individual geometric shapes more formally and to compare their properties. At this level, students can employ a coordinate system to describe the location and side lengths of figures and the distance between figures. Students can then use these data to classify shapes according to various criteria. In the middle grades, students make explicit conjectures and then test their conjectures, searching for counterexamples. For conjectures that appear to be true, students may offer a justification. In subsequent mathematics courses, students continue to offer justifications for conclusions, culminating in mathematical proofs of various forms such as two-column proofs, indirect proofs, and paragraph proofs.

At no grade level are geometric topics restricted to the study of formal geometry.

As in the lower grades, spatial visualization continues to be an important topic in grades 6–8. Middle-grades students come to recognize congruent figures in different orientations and from different perspectives. Transformations such as slides, flips, and turns may be described and quantified by means of a coordinate system. Symmetry may be analytically studied on a coordinate system as students compare the distances of corresponding points from the line of symmetry. In the higher grades, dilations and rotational symmetry are studied as means for describing relationships among geometric figures.

The study of the relationship between two- and three-dimensional figures is begun in earnest in the middle grades. Students make nets and fold them up to determine if they form an appropriate geometric solid. They draw two-dimensional diagrams that represent three-dimensional shapes, and they construct three-dimensional figures that are represented by two-dimensional drawings, diagrams, and pictures. These experiences are invaluable to students when they later study conic sections, areas under curves, and volumes of solids formed by rotating figures.

In the primary grades, geometric models may be used to introduce fractions and the concepts of part and whole. Geometric models such as the number line, grids, and arrays of squares—whether using whole numbers or fractions—enhance multiplication concepts. Students in elementary school learn to display data in a variety of ways, including bar and line graphs. When students interpret bar graphs, they are using their understanding of size to communicate about number.

In the middle grades, geometric models become increasingly important in simulating real-world events. Students may gather data from these simulations to analyze real-life applications. Similar figures are a means to demonstrate proportionality. Proportional relationships, such as percents, may be advantageously represented by geometric models to help students understand multiplicative relationships in contrast to previously learned additive ones. As students continue to become more proficient in data representation, they study pie charts and circle graphs. Finally, students in the middle grades begin to use coordinate geometry to study transformations and to compare figures.

At the secondary level, students use geometry in more-sophisticated mathematics settings. Their study of statistics includes the bell-shaped curve and related distortions of it. Coordinate geometry is further refined and extended to conic sections, three dimensions, and polar coordinates. Geometric models are used to solve problems that involve linear programming and probability. In calculus, geometric concepts are employed to determine areas under curves, shapes and volumes of rotating solids, and changes in slopes of lines.

Experiences with geometry in the middle grades should facilitate crossing the bridge from foundational and informal experiences with geometry concepts and reasoning at the elementary level to the formal and deductive experiences at the secondary level. Each chapter in this book emphasizes reasoning and thinking skills that are necessary to making that crossing: reasoning about the characteristics and properties of shapes, the use of multiple representations as tools for representation and analysis, the investigation and interpretation of geometric objects through transformations and symmetry, and reasoning and visualizing about spatial relationships. The activities and information in this book were designed to facilitate students' progression toward mathematical power and literacy.

GRADES 6–8

NAVIGATING *through* GEOMETRY

Appendix

Blackline Masters and Solutions

Geodee's Sorting Scheme

Name ______________________________

Geodee sorted a set of shapes into two categories. She placed them as shown below.

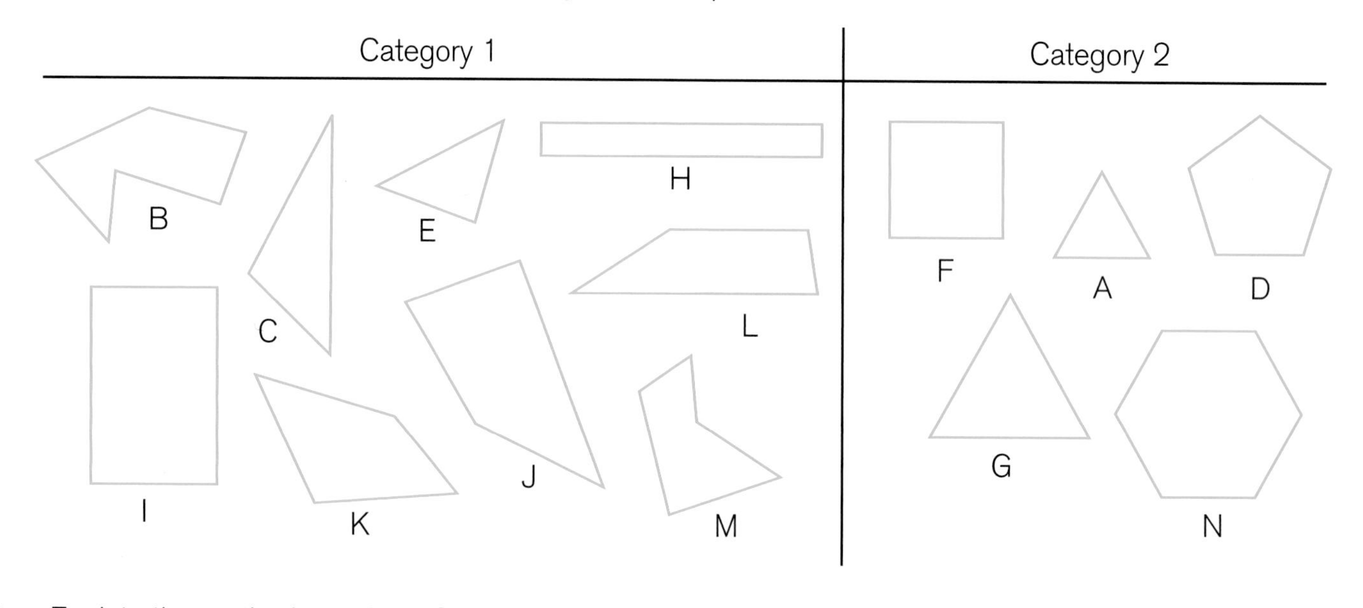

1. Explain the method you think Geodee used to place the shapes in each category.

____

____

____

____

2. How would you define her categories?

____

____

3. After she finished placing the shapes, Geodee realized she had forgotten one. In which category should Geodee place shape O?

____

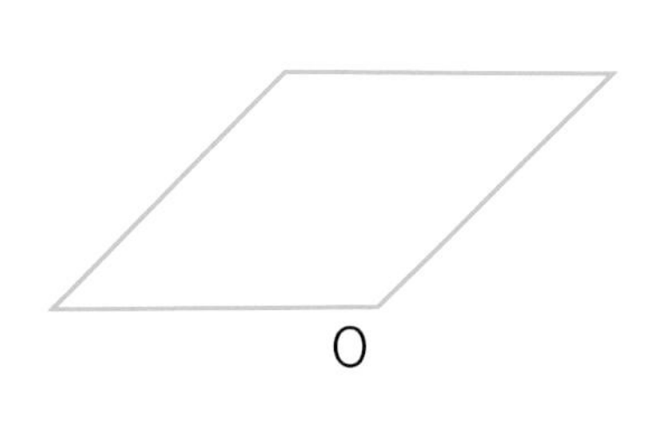

Explain why you think so.

____

____

____

Exploring Triangles

Name ______________________________

1. Using dynamic geometry software, construct a triangle *ABC*.

2. Place the pointer on point *C*, and move the point closer to side *AB*.

   (*a*) Describe what happens to the measure of angle *BCA* as angle *BCA* moves closer to side *AB*.

   ____

   ____

   ____

   (*b*) Describe what happens to the lengths of sides *AC*, *AB*, and *BC*. ______________________________

   ____

   ____

   ____

   ____

3. Move point *C* until it lies on side *AB*.

   (*a*) What happens to angle *BCA*, and why? ______________________________

   ____

   ____

   ____

   (*b*) What happens to the sides of the triangle? ______________________________

   ____

   ____

   ____

Exploring Triangles (continued)

Name ______________________

4. Move point *C* so that sides *AC* and *BC* are of equal length.

   (*a*) What do you notice about the angles of the triangle? ______________________

   ____

   ____

   ____

   (*b*) What do you think would happen if you made side *AB* equal in length to side *BC*?

   ____

   ____

   ____

   ____

5. Write a conjecture about the relationship between the lengths of the sides and the measures of the angles of any triangle.

   ____

   ____

   ____

   ____

   ____

   ____

   ____

   ____

Congruent and Similar Shapes

Which pairs of figures are the same shape?

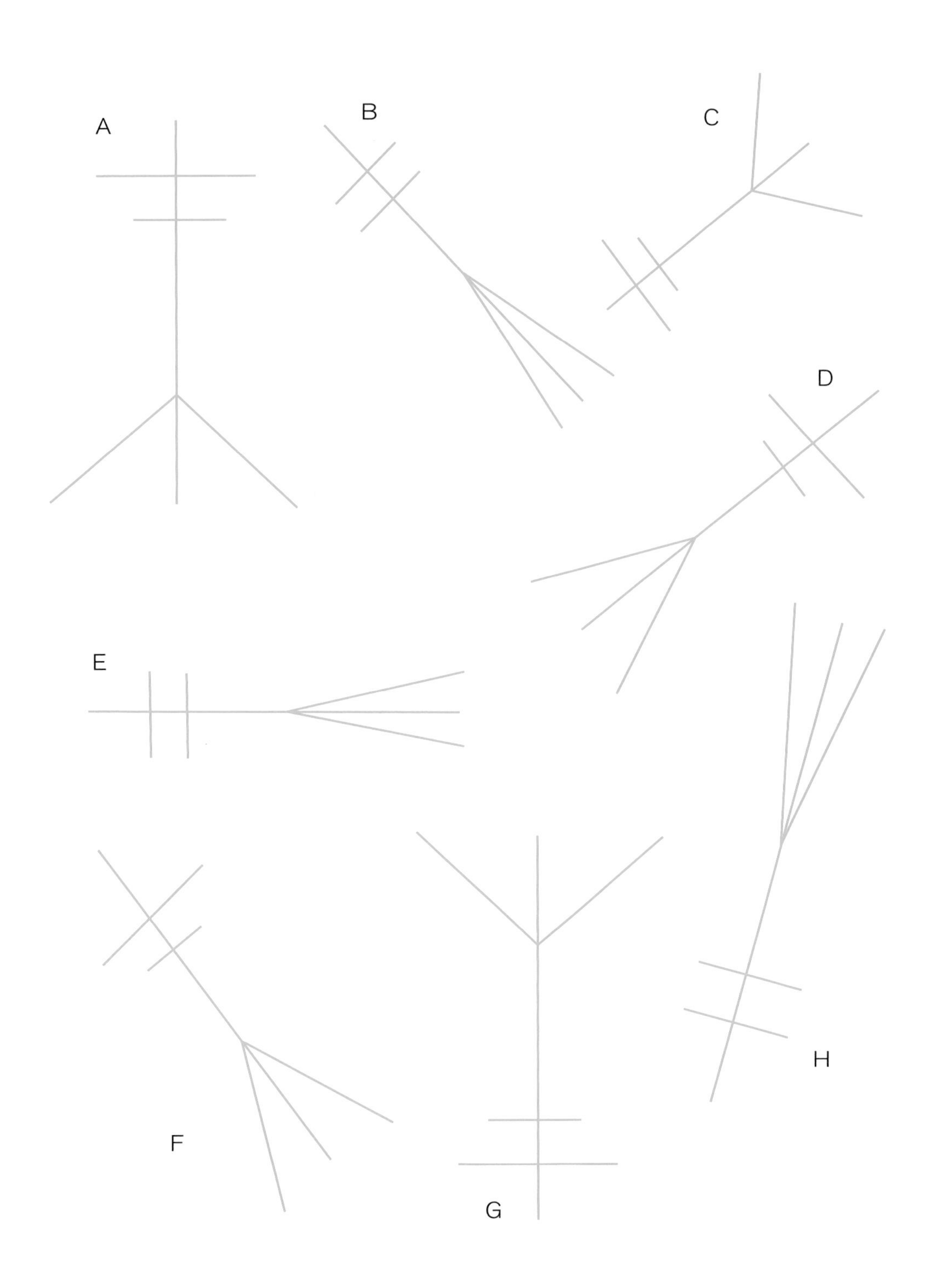

This activity has been adapted from O'Daffer and Clemens (1992, pp. 250–51).

Comparing Elephants

Compare each elephant with elephant A. Describe the differences and similarities between elephant A and each of the other elephants.

This activity has been adapted from CRDG (n.d., unit 5).

Exploring Similar Figures

Name ______________________________

Kristin had to enlarge figure *EFGHIJK.* She worked very hard. Just as she completed the enlargement, she spilled her fruit punch on her homework paper. Help Kristin complete the enlargement. Describe your method.

____

____

____

____

____

____

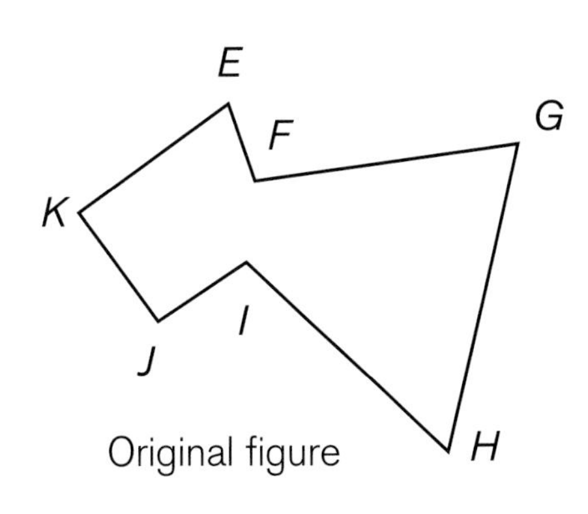

Original figure

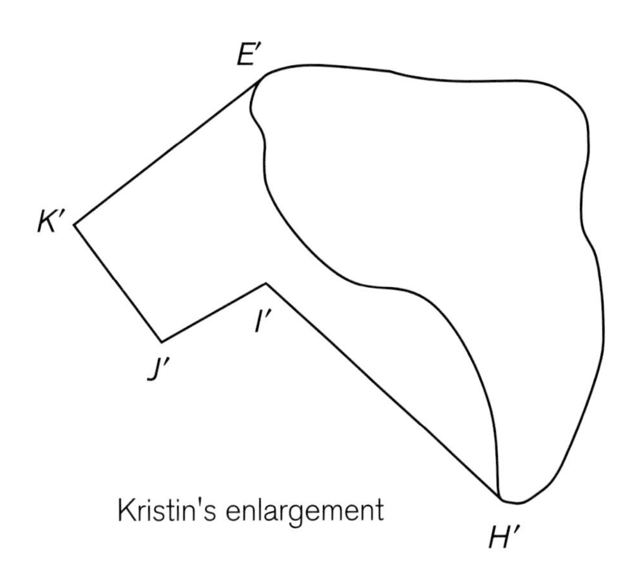

Kristin's enlargement

Dilating Figures

Name ______________________________

Sheri found a good way to change the size of a figure by doing a dilation. She is dilating pentagon *ABCDE* 200 percent to get pentagon *A′B′C′D′E′*.

1. Describe how you think Sheri is making her drawing.

____

____

____

____

____

____

2. Complete the dilation, using *T* as the center.

3. What do you notice about the two figures?

____

____

____

____

____

____

____

____

____

____

____

____

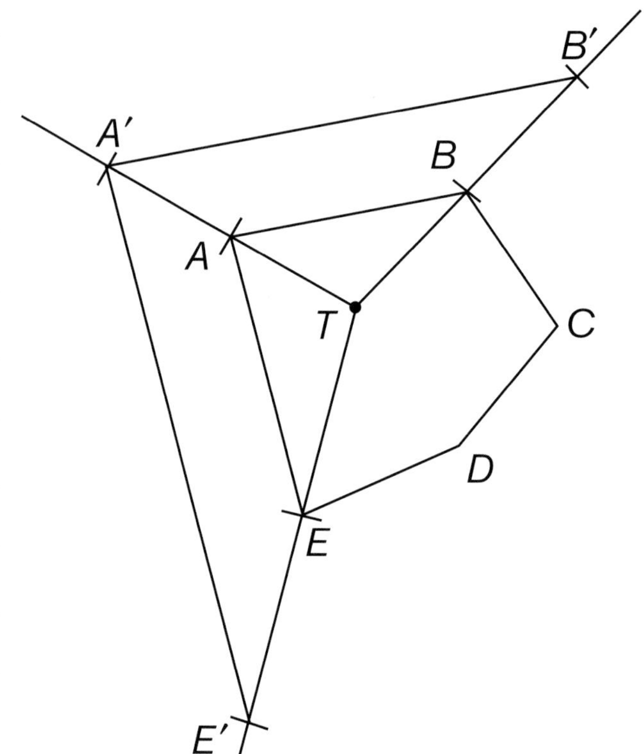

Using Venn Diagrams to Reason about Shapes

Name ______________________________

1. Using the Venn diagram below, label sets A and B with any of the following types of triangles: acute, equilateral, isosceles, obtuse, right, or scalene.

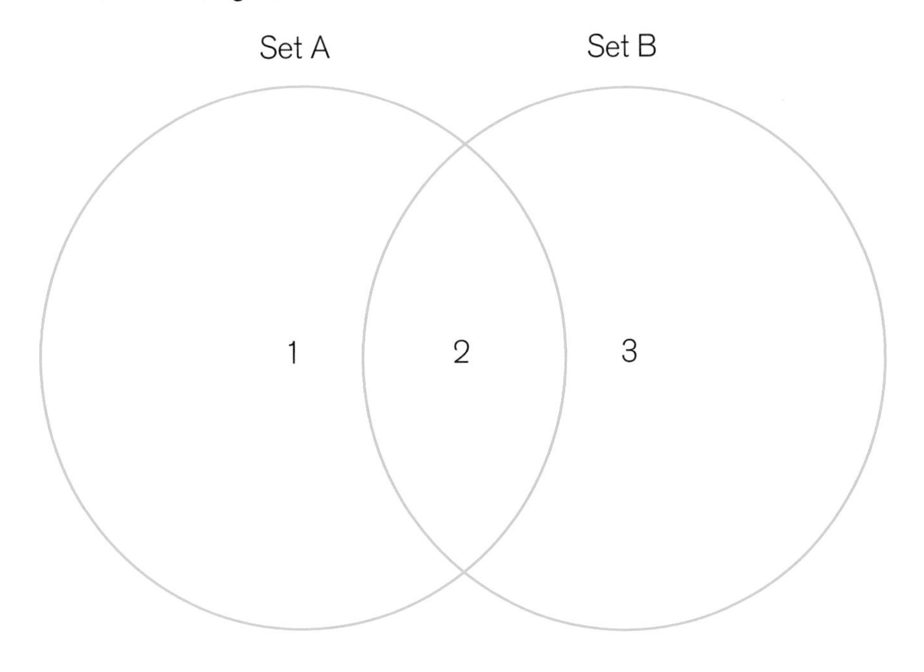

2. Explain why you labeled the Venn diagram as you did.

____

____

____

3. Draw examples of appropriate triangles in each of regions 1, 2, and 3.

4. What are the relationships among these sets of triangles?

____

____

____

____

____

____

Midsegments of Triangles

Name ______________________

1. Construct five triangles using interactive geometry technology or by hand. Label the sides *a, b,* and *c.*

2. Locate the midpoint of two sides of each triangle. In each triangle draw the line connecting the two midpoints. You have just constructed a midsegment of the triangle.

3. Measure and record the lengths of the midsegments and the lengths of all sides of each triangle. Use the table below to record your data.

4. Look for patterns in your data and describe any that emerge.

____

____

____

____

____

____

Triangle Number	Length of Side *a*	Length of Side *b*	Length of Side *c*	Length of Midsegment

Reasoning about the Pythagorean Relationship

Name ______________________________

1. Using the distance between adjacent points of intersection on grid paper as units, draw a right triangle with sides measuring 3, 4, and 5 units. Let the legs be 3 units and 4 units and lie on grid lines. Place a second sheet of grid paper along the hypotenuse to verify that it has a length of 5 units.

2. Draw squares, using the sides of the triangle as sides of the squares, as illustrated at the right. Find the area of each square, and enter the information in the table below.

3. Similarly, draw a right triangle with sides measuring 5, 12, and 13 units. Draw squares on the sides of the new triangle. Find the area of each square, and enter the information in the table.

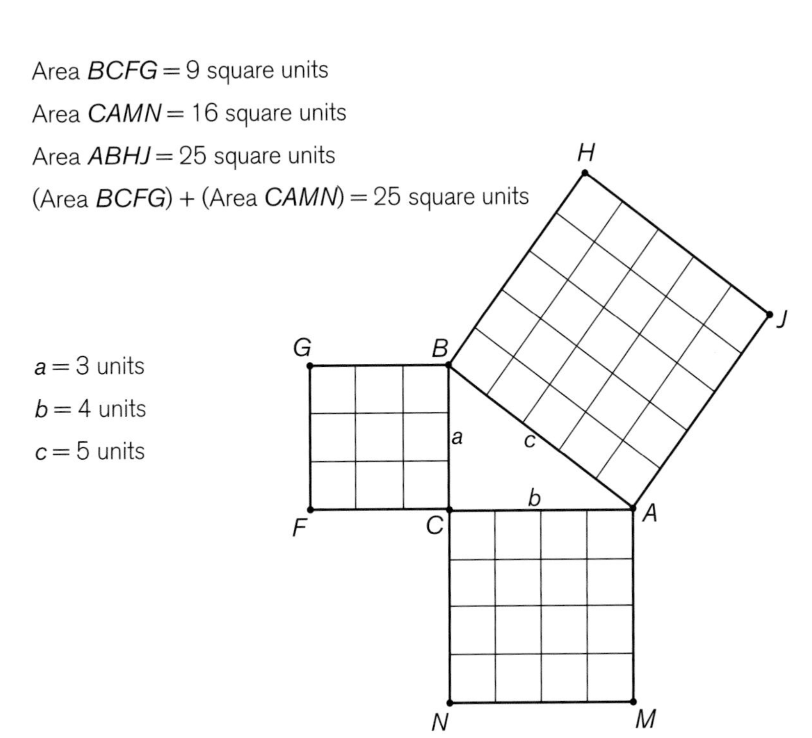

Triangle Number	Length of Side *a* (Units)	Length of Side *b* (Units)	Length of Side *c* (Units)	Area of Square with Length *a* (Square Units)	Area of Square with Length *b* (Square Units)	Area of Square with Length *c* (Square Units)
1	3	4	5	9	16	25
2	5	12	13			
3						
4						

4. Draw other triangles, as instructed by your teacher, and record the appropriate data in the table. Look for patterns in the data. Describe any relationships among the parts of the figure.

____

____

____

____

Constructing Geometric Figures in Coordinate Space

Name ____________________________

Draw the coordinate plane on grid paper, construct the figures described in the left-hand column of the chart below, and write the coordinates of the figures in the right-hand column.

Description of Figures	Coordinates of Figures
1. A square with sides of 3 units	
2. A rectangle with dimensions 2 units by 4 units	
3. A square with sides of 5 units and one vertex at (−1, −1)	
4. At least four other squares meeting the conditions in the previous description	
5. A rectangle with a vertex at (1, 2) and dimensions 3 units by 4 units	
6. At least four other rectangles meeting the conditions in the previous description	
7. A rectangle whose perimeter is between 12 units and 16 units	
8. Two other rectangles meeting the conditions in the previous description	
9. A square with vertices at (3, 4) and (3, 8)	
10. Two other squares meeting the conditions in the previous description	
11. A square with a perimeter between 16 units and 20 units and with a vertex at (1, 2)	
12. A right triangle with vertices at (0, 0) and (0, −6)	
13. An acute triangle with vertices at the coordinates given in the previous description	
14. A right triangle with the vertex of the right angle at (−5, −8) and having legs measuring 4 units and 2 1/2 units	

Exploring Lines, Midpoints, and Triangles Using Coordinate Geometry

Name ______________________________

Use the graphs to plot the points, as directed.

Graphs, Coordinates, and Lines

1. Plot the following points on the first graph: *A:* (0, 2), *B:* (3, 5), *C:* (5, 7), *D:* (6, 8), *E:* (8, 9), and *F:* (9, 11).

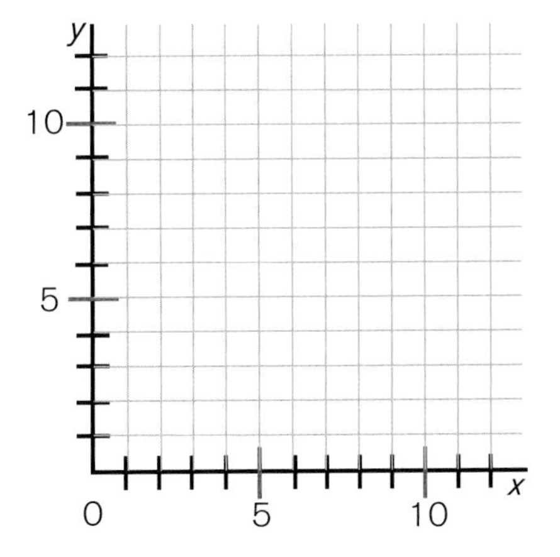

2. Which one of the six points in your graph seems to be out of place? ______________________________

____

3. The coordinates of the other five points have something in common that is not shared by the sixth point. What is it?

____

4. Plot the following points on the second graph: *A:* (4, 0), *B:* (4, 1), *C:* (4, 3), *D:* (4, 7), and *E:* (4, 10).

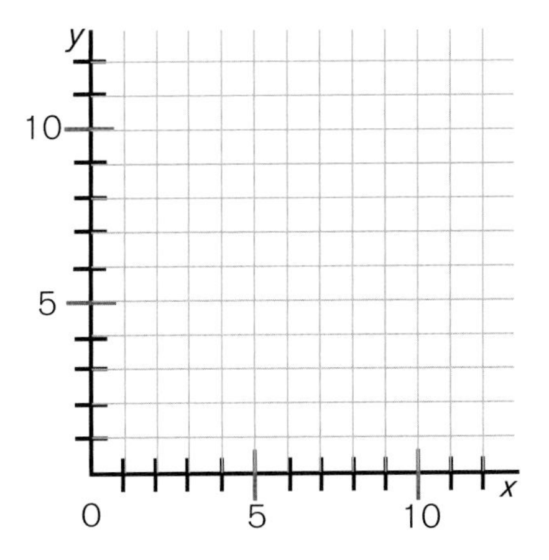

5. What do you notice about the five points in your graph?

____

6. What do the coordinates of the five points have in common?

____

____

Graphs, Coordinates, and Midpoints

1. Plot the following points on the third graph: *A:* (1, 6), *B:* (5, 2), *C:* (6, 7), *D:* (12, 9).

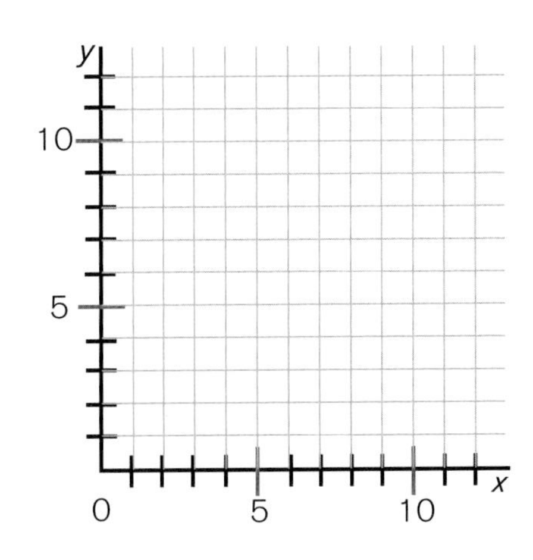

2. Find the point midway between points *A* and *B,* and label it *M.* What are the coordinates of point *M*? ______________

3. How are the coordinates of point *M* related to the coordinates of points *A* and *B*? ______________________

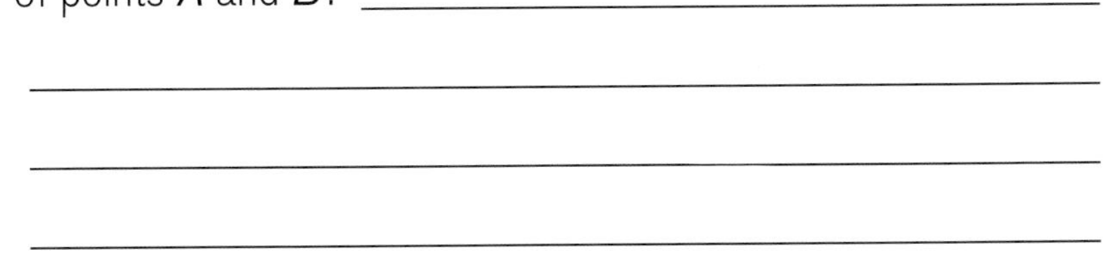

Exploring Lines, Midpoints, and Triangles Using Coordinate Geometry (continued)

Name ______________________________

4. Find the point midway between points *C* and *D*, and label it *N*. What are the coordinates of point *N*?

____

5. Are the coordinates of point *N* related to the coordinates of points *C* and *D* in the same way that the coordinates of point *M* are related to the coordinates of points *A* and *B*? ______________

   Explain your answer. ______________________________

____

____

____

____

Graphs, Coordinates, and Triangles

1. Plot the following points on the fourth graph. Connect the vertices with line segments to form the three triangles named.

   Triangle 1: (2, 1), (5, 2), (3, 6)

   Triangle 2: (5, 5), (8, 6), (6, 10)

   Triangle 3: (9, 0), (12, 1), (10, 5)

y
10
5
0 5 10 x

2. In what ways are the three triangles you have drawn alike?

____

____

____

3. The *x*-coordinates of the vertices of triangle 2 are three more than the *x*-coordinates of the corresponding vertices of triangle 1. How do the *y*-coordinates of the vertices of triangle 2 compare with the *y*-coordinates of the corresponding vertices of triangle 1? ______________

____

4. How are the coordinates of the vertices of triangle 3 related to the coordinates of the corresponding vertices of triangle 1?

____

____

____

Exploring Lines, Midpoints, and Triangles Using Coordinate Geometry (continued)

Name ______________________________

5. Plot the following points on the fifth graph, and connect them with line segments to form the three triangles named:

   Triangle 1: (4, 0), (6, 6), (2, 4)

   Triangle 2: (8, 0), (12, 12), (4, 8)

   Triangle 3: (2, 0), (3, 3), (1, 2)

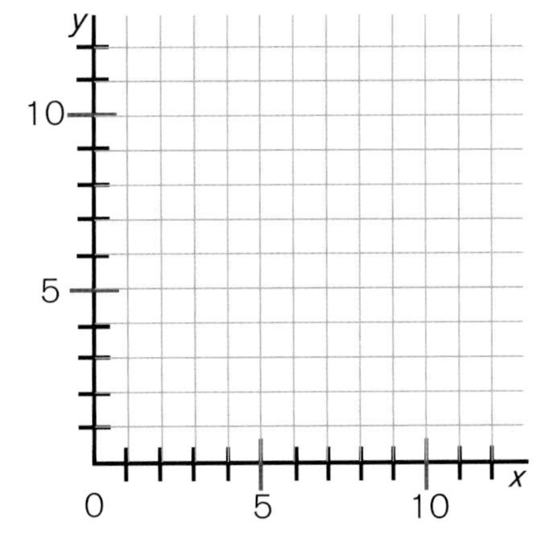

6. In what ways are the three triangles that you have drawn alike? ______________________________

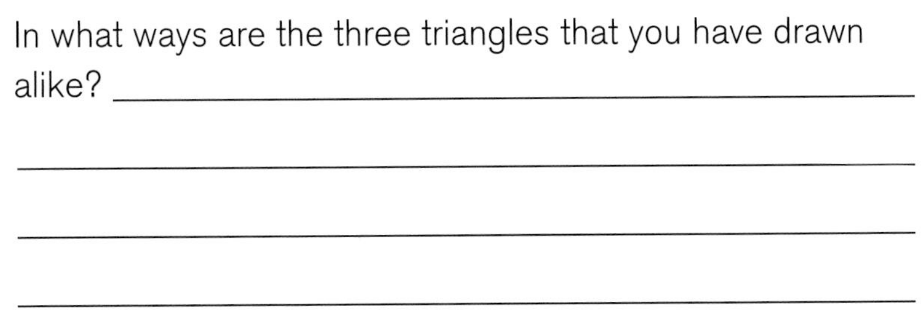

7. How are the coordinates of the vertices of triangle 2 related to the coordinates of the corresponding vertices of triangle 1? ______________________________

8. How are the coordinates of the vertices of triangle 3 related to the coordinates of the corresponding vertices of triangle 1? ______________________________

Similarity and the Coordinate Plane

Name ________________________________

Consider triangle $A'B'C'$ and pentagon $D'E'F'G'H'$, that you have drawn as instructed by your teacher.

1. What do you notice about the points A', B', C', D', E', F', G', and H'?

____

____

____

2. How do the distances OA' and OA compare?

____

____

   How do the distances OB' and OB compare?

____

____

3. Do the two triangles ABC and $A'B'C'$ have the same shape? ____________________

   Do the two pentagons have the same shape? ____________________________

4. Make a conjecture about what the results in question 2 would be if all coordinates were multiplied by 3, by 4, or by 1/5.

____

____

____

____

   Check your conjecture with drawings on grid paper.

5. Draw on grid paper a picture of your own choosing. Draw the figure resulting from multiplying all the coordinates in your drawing by 2.

This activity has been adapted from O'Daffer and Clemens (1992, p. 239).

Exploring the Slopes of Parallel and Perpendicular Lines

Name ______________________________

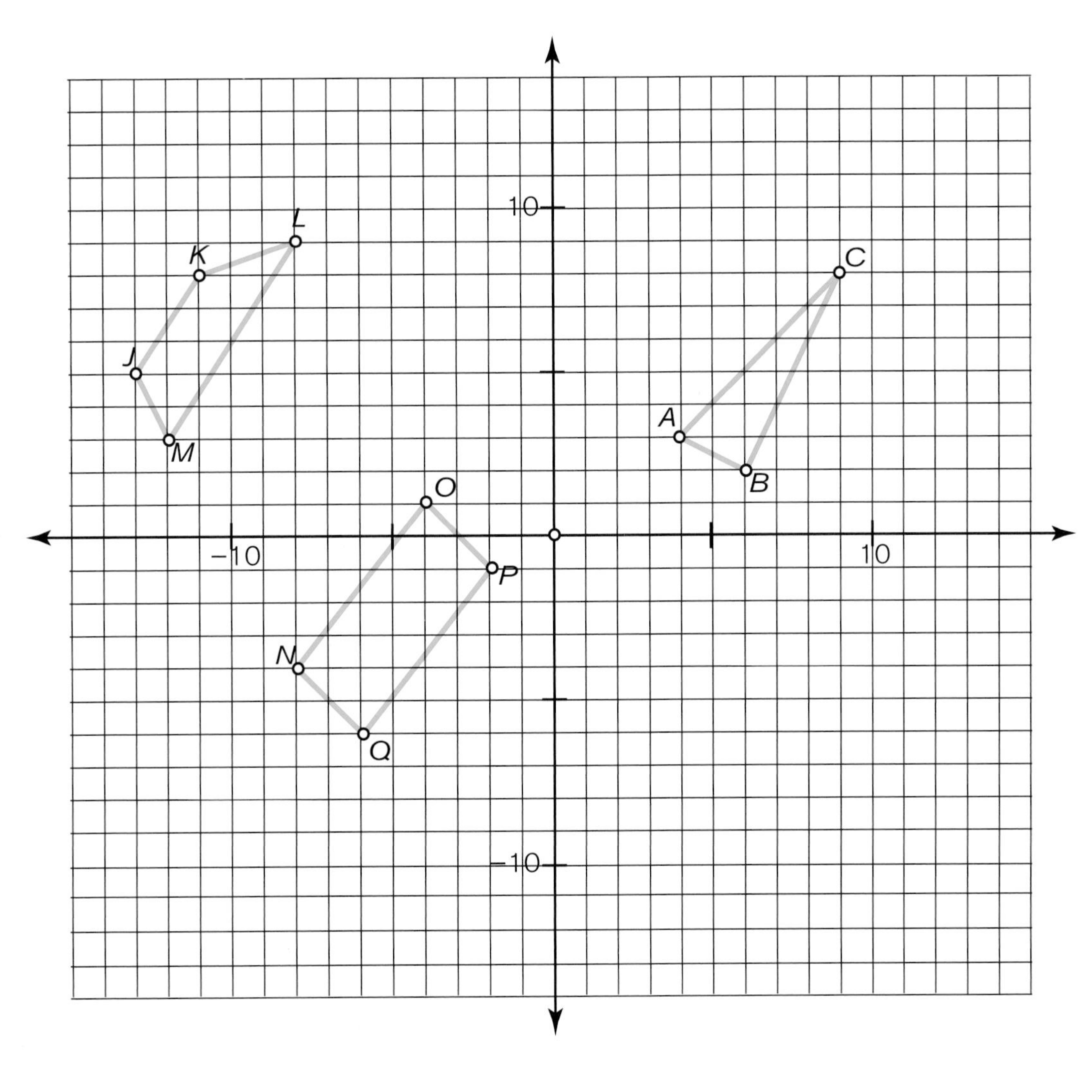

1. Use the space below to determine the slopes of each of the sides of the polygons in the graph above. Record your results here.______________________________

____

____

Name ______________________________

2. Compare the slopes of the line segments that form each figure.

   (*a*) Which sides have the same slope? ______________________________

   ____

   (*b*) In each figure, what is the relationship between the sides that have the same slope? __________

   ____

   ____

   ____

   (*c*) What is the primary difference between *JKLM* and *NOPQ*? ______________________________

   ____

   ____

   ____

   (*d*) What is the relationship between the slopes of segments *AB* and *BC*? ______________________________

   ____

   What is the measure of $\angle ABC$? ______________________________

3. Write a conjecture about the slopes of parallel and perpendicular lines. ______________________________

   ____

   ____

   ____

   ____

   Support your conjecture with additional figures illustrating the relationship.

Reflection of Images

Name ___________________________

1. Draw the image of each of the four figures below when the original shape has been reflected across the *y*-axis.

A.

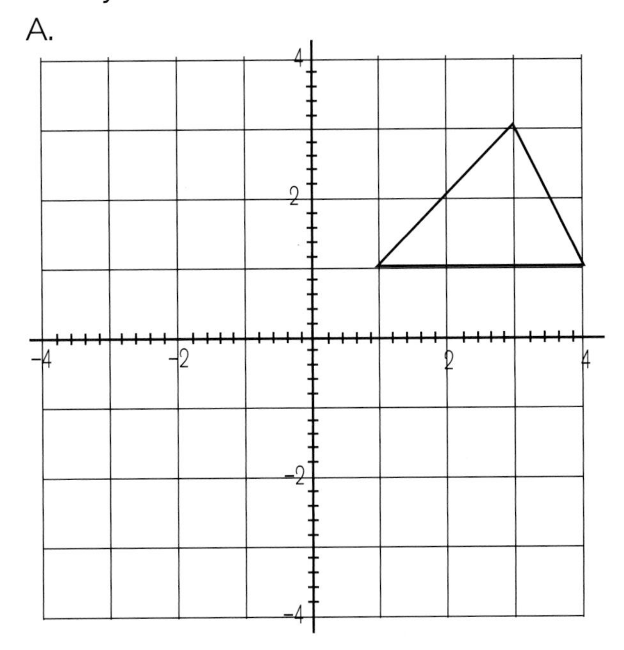

B.

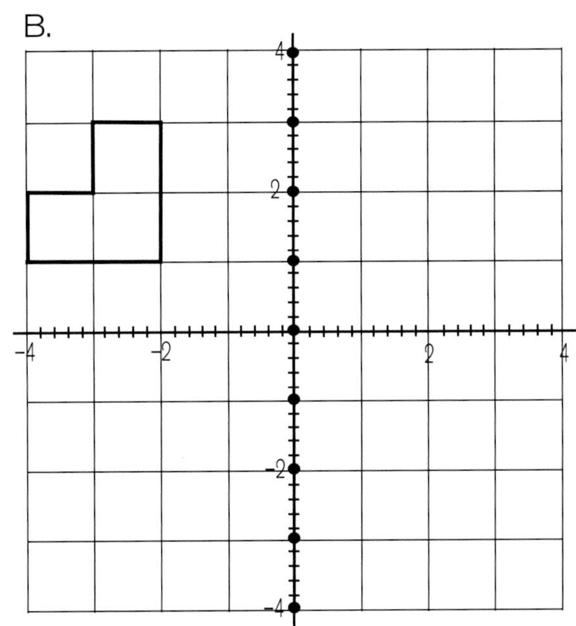

C.

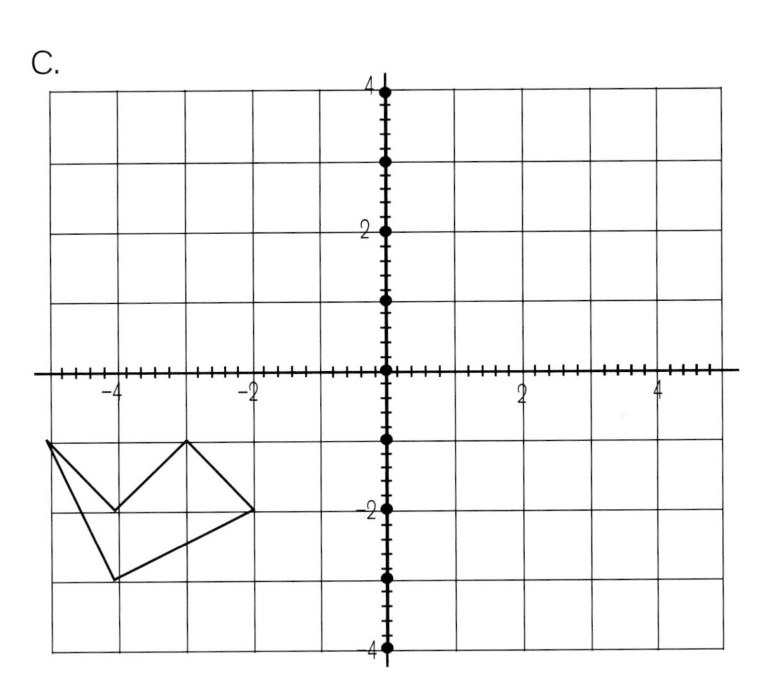

D.

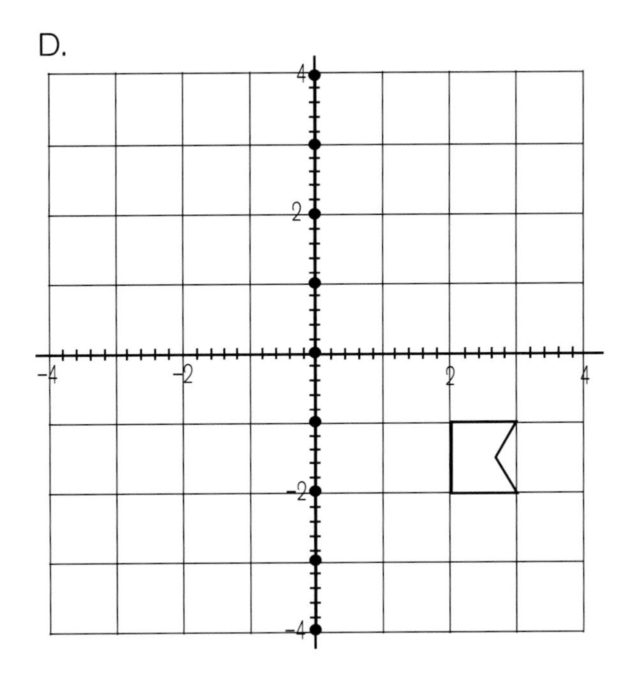

2. What image results when each of the original figures has been reflected across the *x*-axis? Draw the reflected figures.

Translations, Reflections, and Rotations

Name ______________________________

Complete the following steps for a translation:

1. Draw an arrow (vector) on one sheet of translucent paper. Use less than half the sheet to allow enough room for the translated figure. Make sure the point and the arrow are dark enough to be seen through a second sheet of paper.

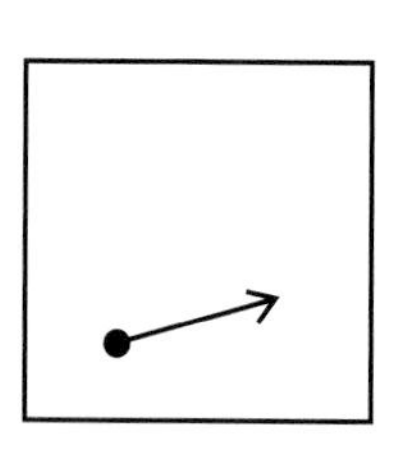

2. Draw an original shape on the sheet of paper.

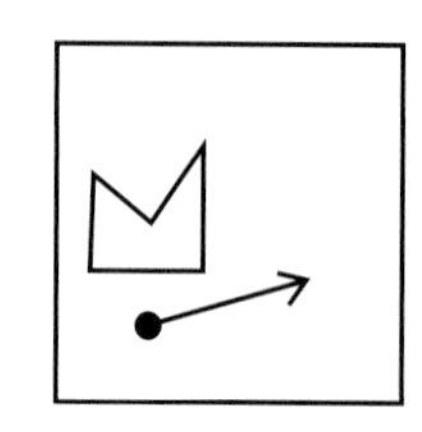

3. Place a second translucent sheet over the first sheet. On sheet 2, trace the endpoint of the arrow and draw a line that extends beyond the endpoint and head of the original arrow. Without moving sheet 2, trace the original figure.

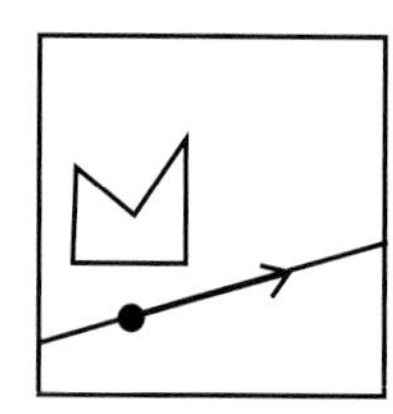

4. Place the second sheet under the original sheet. Align the line on the second sheet and the arrow (vector) on the first sheet. Slide sheet 2 until the point you drew on the line is under the tip of the arrow.

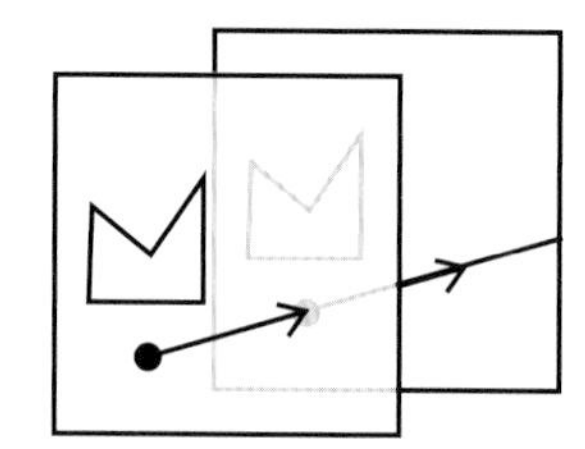

5. Trace the image from sheet 2 onto sheet 1. Label the original figure and the translated figure.

This activity has been adapted from the Shapes and Measurement Module (pp. 198–201), developed by the Reconceptualizing Mathematics Project, Center for Research in Mathematics and Science Education, San Diego State University.

Translations, Reflections, and Rotations (continued)

Name ______________________________

Complete the following steps for a reflection:

1. Draw a shape using less than half a sheet of translucent paper. Draw a line of reflection so that your paper is divided into two areas. Draw a heavy point on the line to use as a reference point.

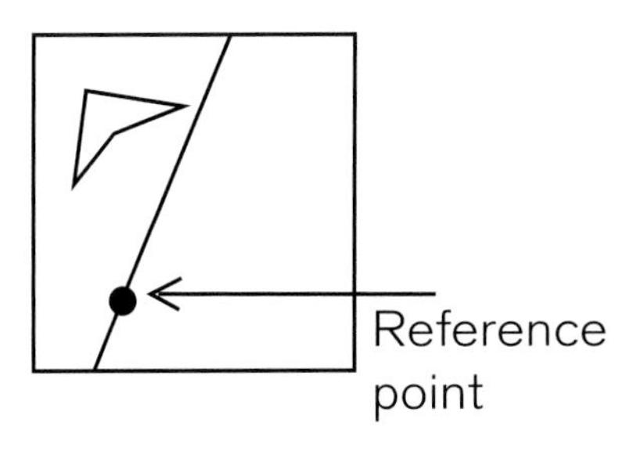

2. Place a second sheet of translucent paper on top of the original, and trace the figure, the line of reflection, and the reference point.

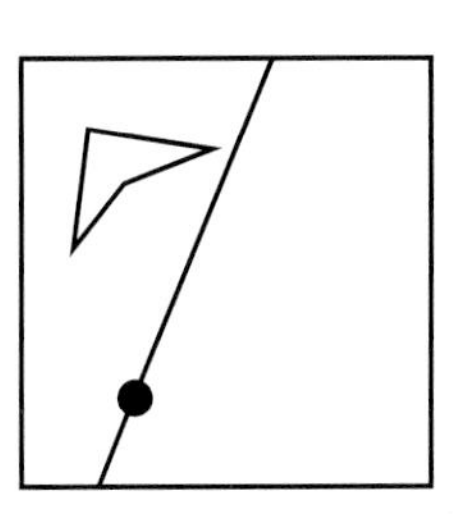

3. Flip sheet 2 over, and put it under the original sheet. Align the lines of reflection and the reference points.

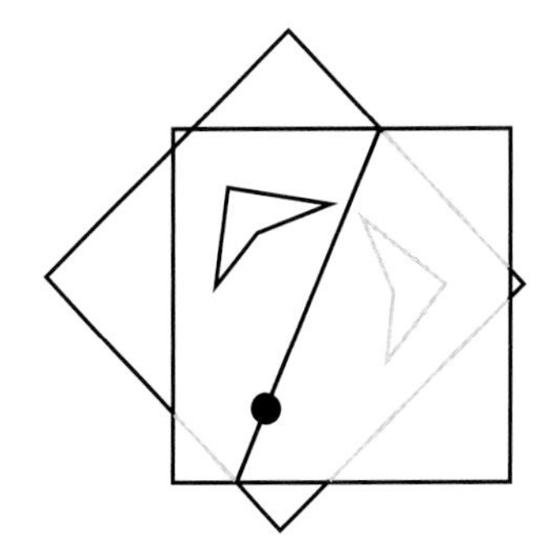

4. Trace the image from sheet 2 onto the original sheet. Label the original figure and the reflected figure.

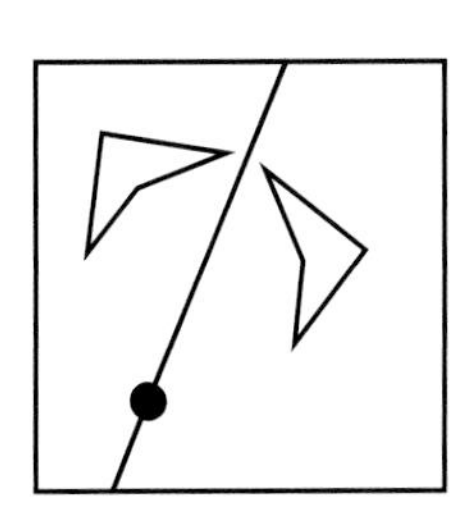

This activity has been adapted from Shapes and Measurement Module (pp. 198–201), developed by the Reconceptualizing Mathematics Project, Center for Research in Mathematics and Science Education, San Diego State University.

Name ______________________________

Complete the following steps for a rotation:

1. Draw a shape on less than half a sheet of translucent paper. On the other half of the paper, draw a point to serve as the center of rotation and draw the angle of the rotation, with the center of rotation as the vertex of the angle. Don't make the angle too large, or you will not have enough room on the paper to draw the rotated figure. A clockwise rotation is illustrated.

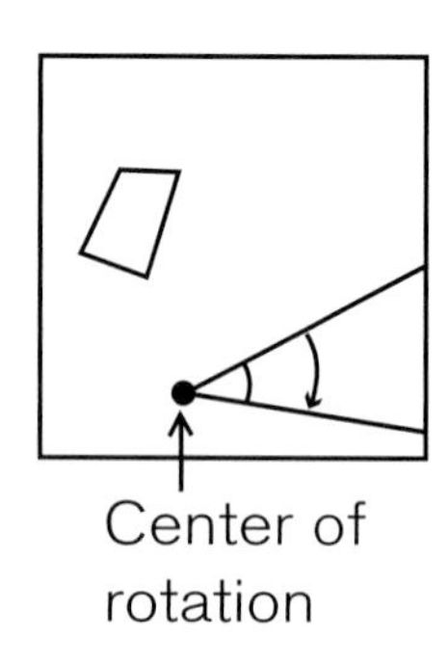

2. Place a second sheet of translucent paper on top of the original. Trace the figure, the vertex (point) of the angle of reflection, and one ray of the angle, as indicated.

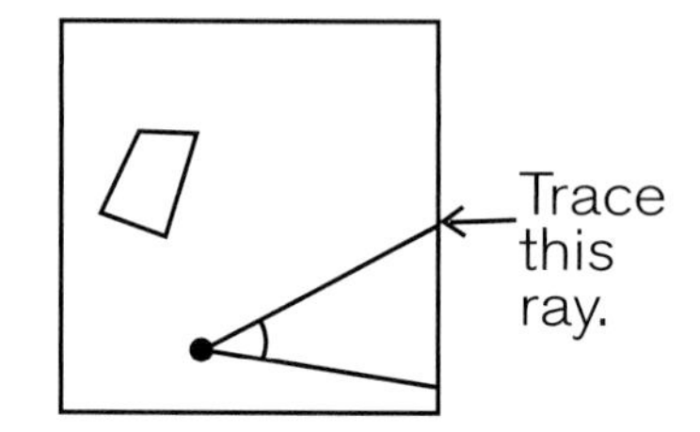

3. Place the second sheet under the original so that everything is aligned. Place your pencil tip on the vertex of the angle (the center of rotation). Turn sheet 2 until the ray is aligned with the other ray of the angle on sheet 1.

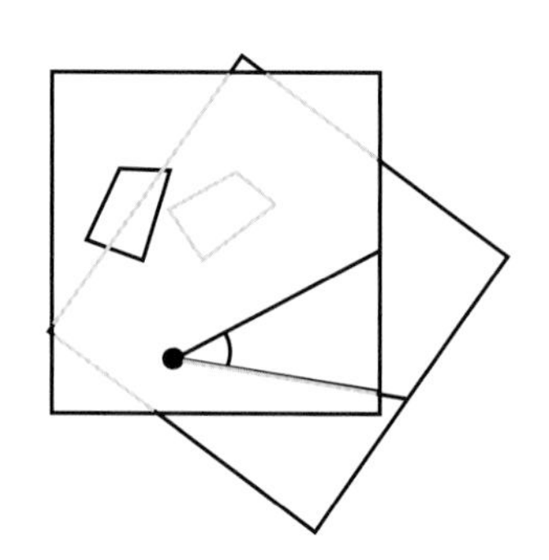

4. Trace the image onto the original sheet. Label the original figure and the rotated figure.

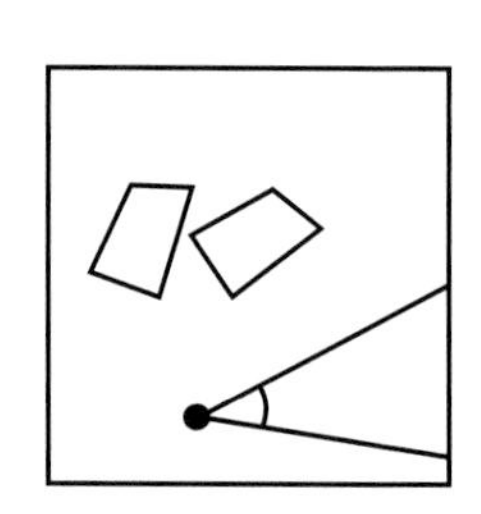

This activity has been adapted from Shapes and Measurement Module (pp. 198–201), developed by the Reconceptualizing Mathematics Project, Center for Research in Mathematics and Science Education, San Diego State University.

Using Scale Factors

Name ___________________________________

To locate the projected and scaled image of a figure, draw rays from the projection point through the vertices of the original figure. Place the vertices of the scaled figure at the points whose distances from *P* are the lengths of the segment from *P* to the corresponding vertices of the original figures multiplied by the scale factor.

1. Using point *P* as the projection point and a scale factor of 2, locate the image *A′B′C′* of triangle *ABC*.

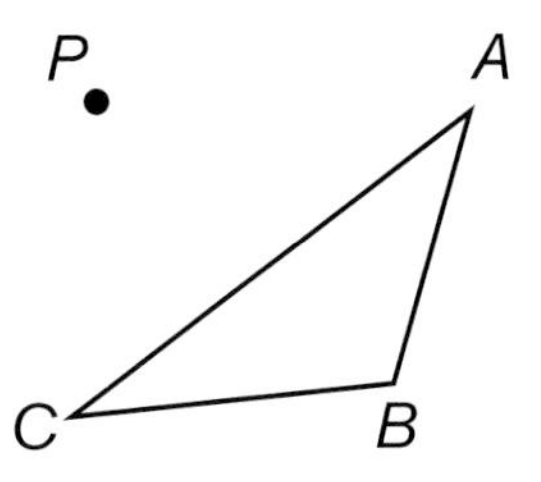

2. Use point *P* as the projection point and a scale factor of 3/4 to find the image *A′B′C′D′* of rectangle *ABCD*.

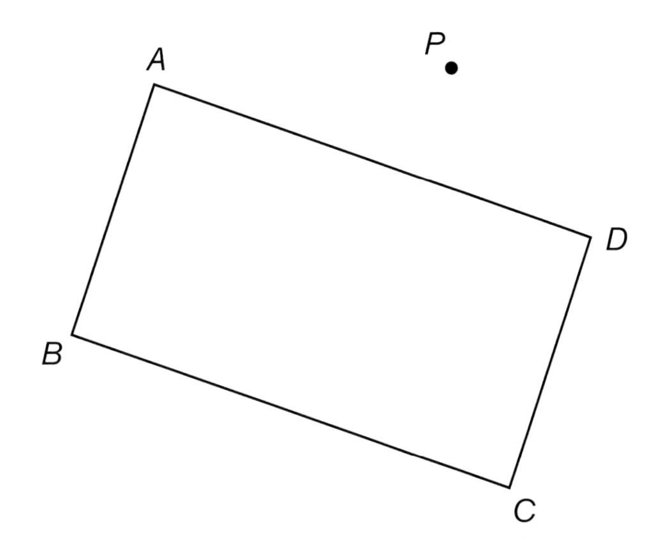

Explorations with Lines of Symmetry

Name ______________________________

1. Look carefully at the four figures below. Is the dashed line in each figure a line of symmetry? Explain your answer for each figure.

A. 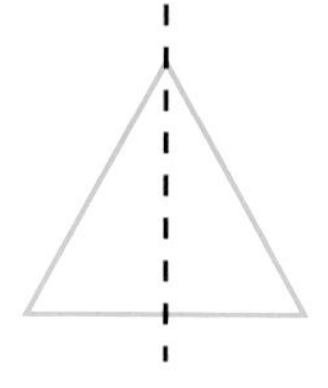

B. 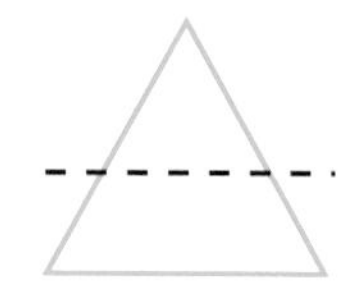

C. 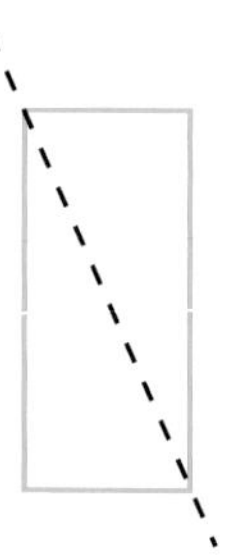

D. 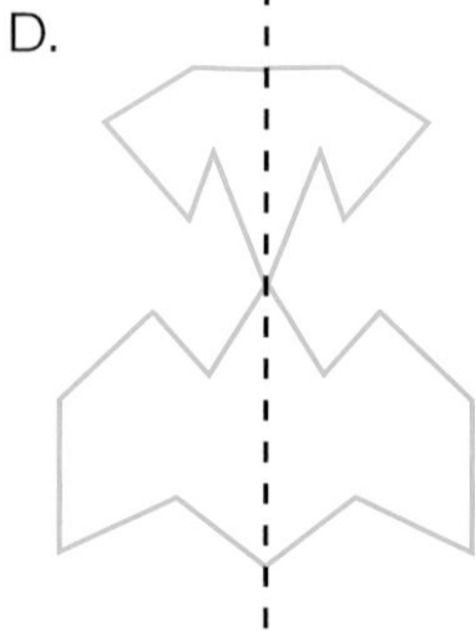

2. Draw two figures that have line symmetry. Draw the lines of symmetry.

Alphabet Symmetry

A	B	C	D
E	F	G	H
I	J	K	L
M	N	O	P
Q	R	S	T
U	V	W	X
Y	Z		

Square Symmetry

Cut line

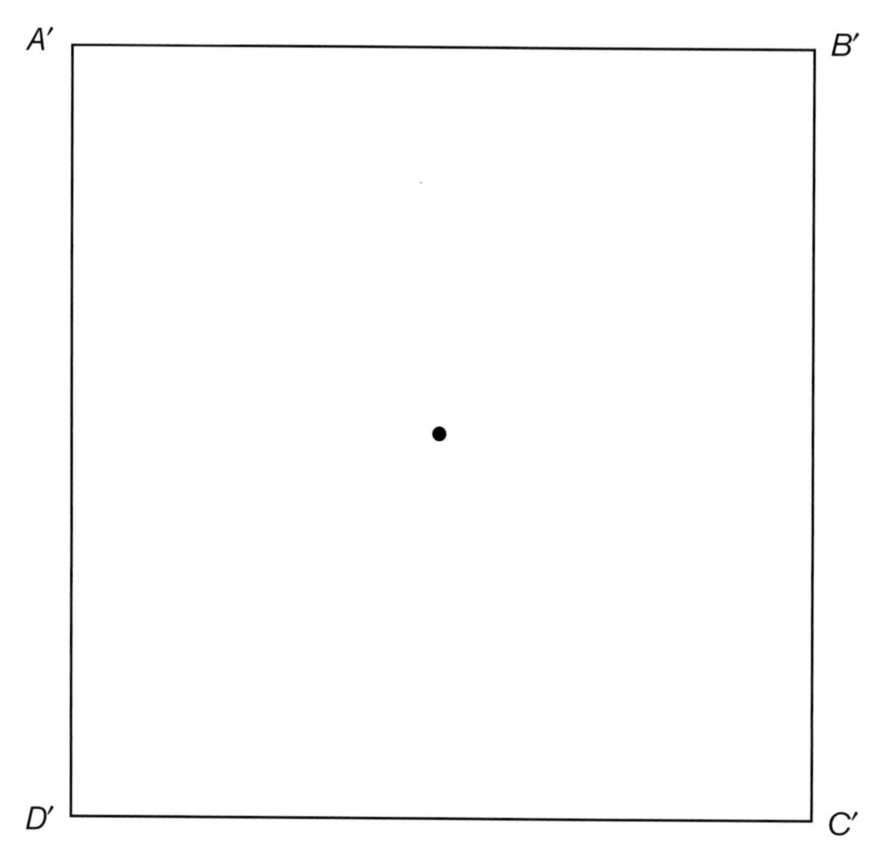

Rotational Symmetry and Regular Polygons

Name ______________________________

1. Draw or trace a large equilateral triangle on one sheet of tracing paper. Locate the center of rotational symmetry by folding two of the angles so that they are bisected. Label the center of rotational symmetry. Place a mark outside the triangle to use as a reference point.
2. Copy the triangle, the center of rotational symmetry, and the reference point onto another piece of tracing paper.
3. Place the copy over the original and hold the tip of your pencil at the center of rotational symmetry. Slowly rotate the copy until the image is overlaying the original triangle. Observe the number of times the image coincides with the original in one complete rotation (360°, or when the reference points coincide).
4. Repeat the steps above with a square and a regular pentagon, hexagon, heptagon, and octagon.
5. Complete the table below, look for patterns in your data, and make a conjecture about the relationship between the number of sides of a regular polygon and the number of rotational symmetries in the polygon.

Rotational Symmetry in Regular Polygons

Number of sides	3	4	5	6	7	8	. . .	n
Number of rotational symmetries	3							
List of rotational symmetries	120° 240° 360°							

Conjecture: ______________________________

____

____

____

Logos and Geometric Properties

Name ____________________________

1. Find logos in a magazine or newspaper or on commercial products such as boxes or printed materials. Cut out or sketch the logos and divide them into the following categories. Some logos may fit more than one category.

   (*a*) Logos that have the following properties:
   - Line symmetry
   - Rotational symmetry
   - Parallel lines

   (*b*) Logos that display the following transformations:
   - Translation
   - Dilation
   - Reflection
   - Rotation

   (*c*) Logos that contain the following shapes:
   - Circle
   - Regular polygon
   - Nonregular polygon
   - Other shapes you like

2. Create your own logo that displays at least two geometric properties. Identify those properties, and describe how the logo demonstrates each one.

____

____

____

____

Isometric Explorations

Name ______________________________

1. Which isometric drawing shows the view from the left front corner of the building represented by the base plan below? _______

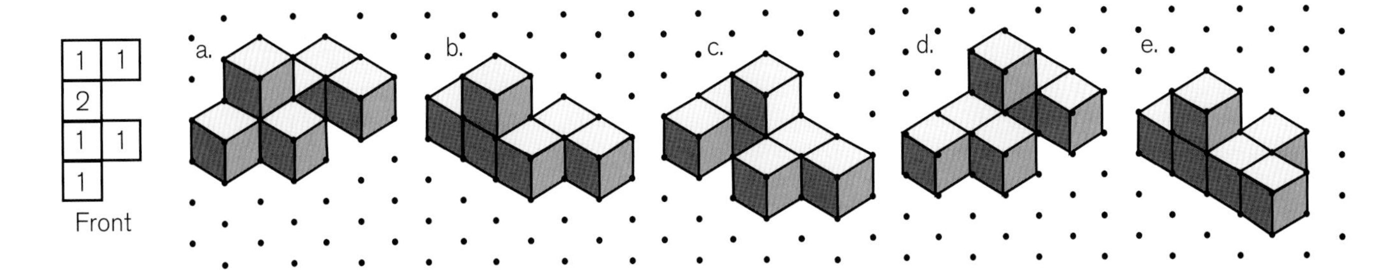

2. Which isometric drawing shows the view from the back left corner of the building represented by the base plan below? _______

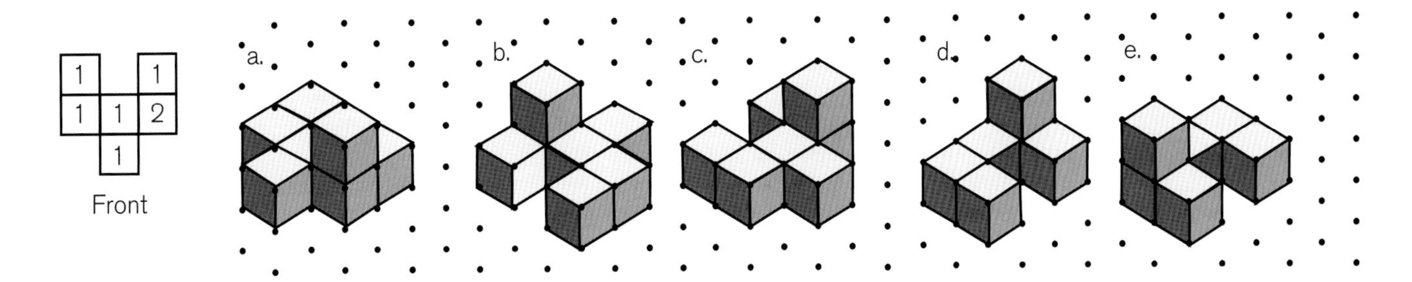

3. Which drawings are views of the building shown below? _______

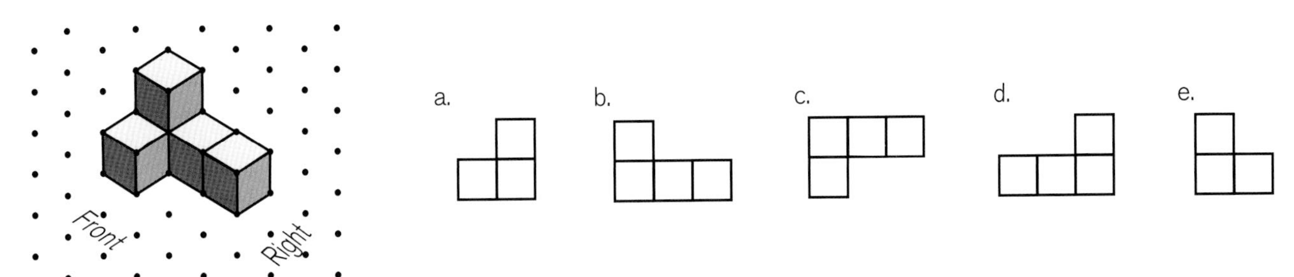

This activity is from *Ruins of Montarek* (Lappan et al. 1996, pp. 87–88). Used with permission.

Isometric Explorations (continued)

Name ______________________________

4. The isometric drawing below shows a building from the front right corner. Which drawing shows the back view of the building? ________

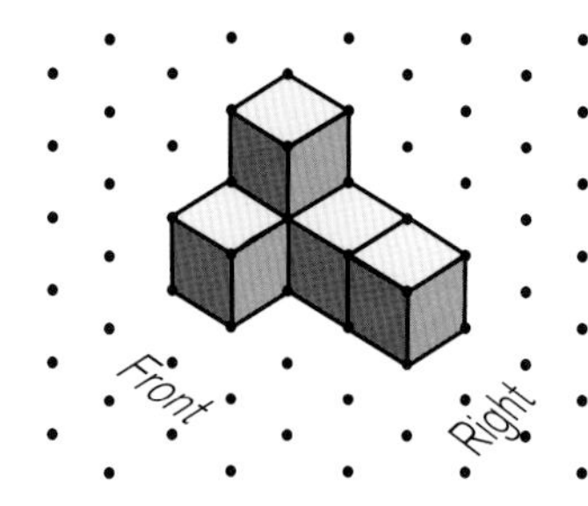

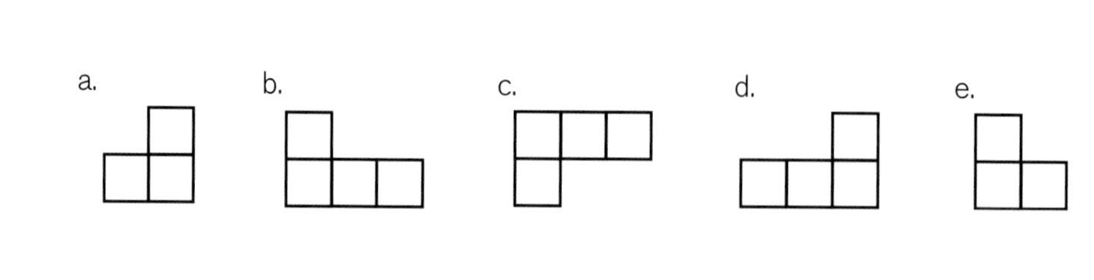

5. The isometric drawing below shows a building from the front right corner. Which drawing shows the right view of the building? ________

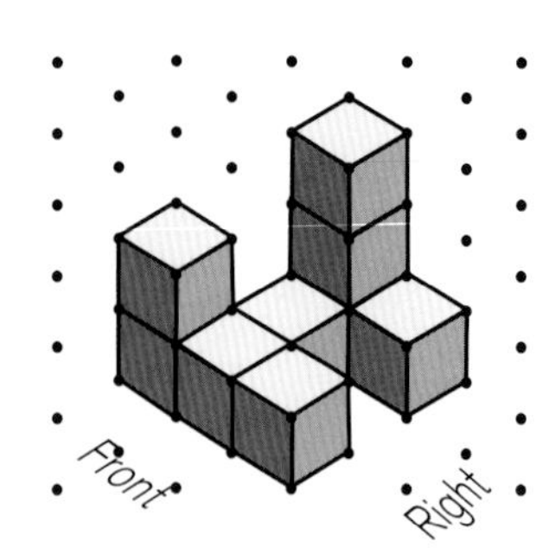

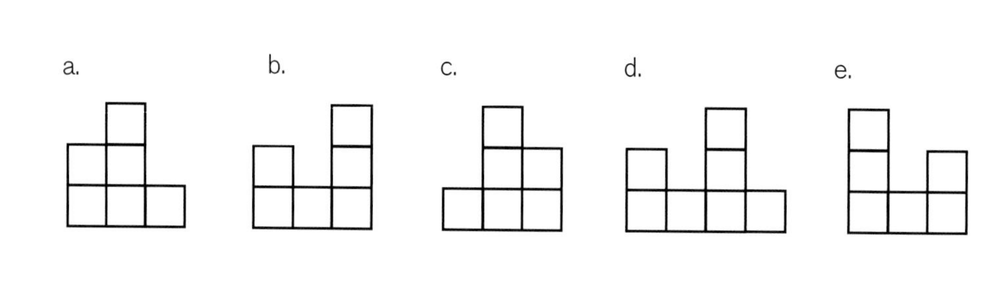

6. Draw the isometric view from the right back corner of the building represented by the base plan below.

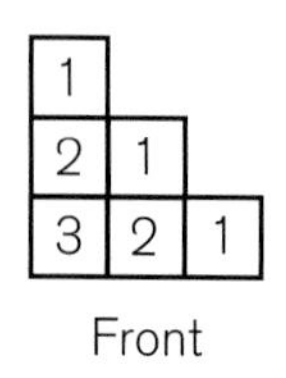

This activity is from *Ruins of Montarek* (Lappan et al. 1996, pp. 87–88). Used with permission.

Cross Sections of Three-Dimensional Shapes

Name ______________________________

For each of the solids below, draw the cross section formed when the plane indicated by the lines intersects the figure.

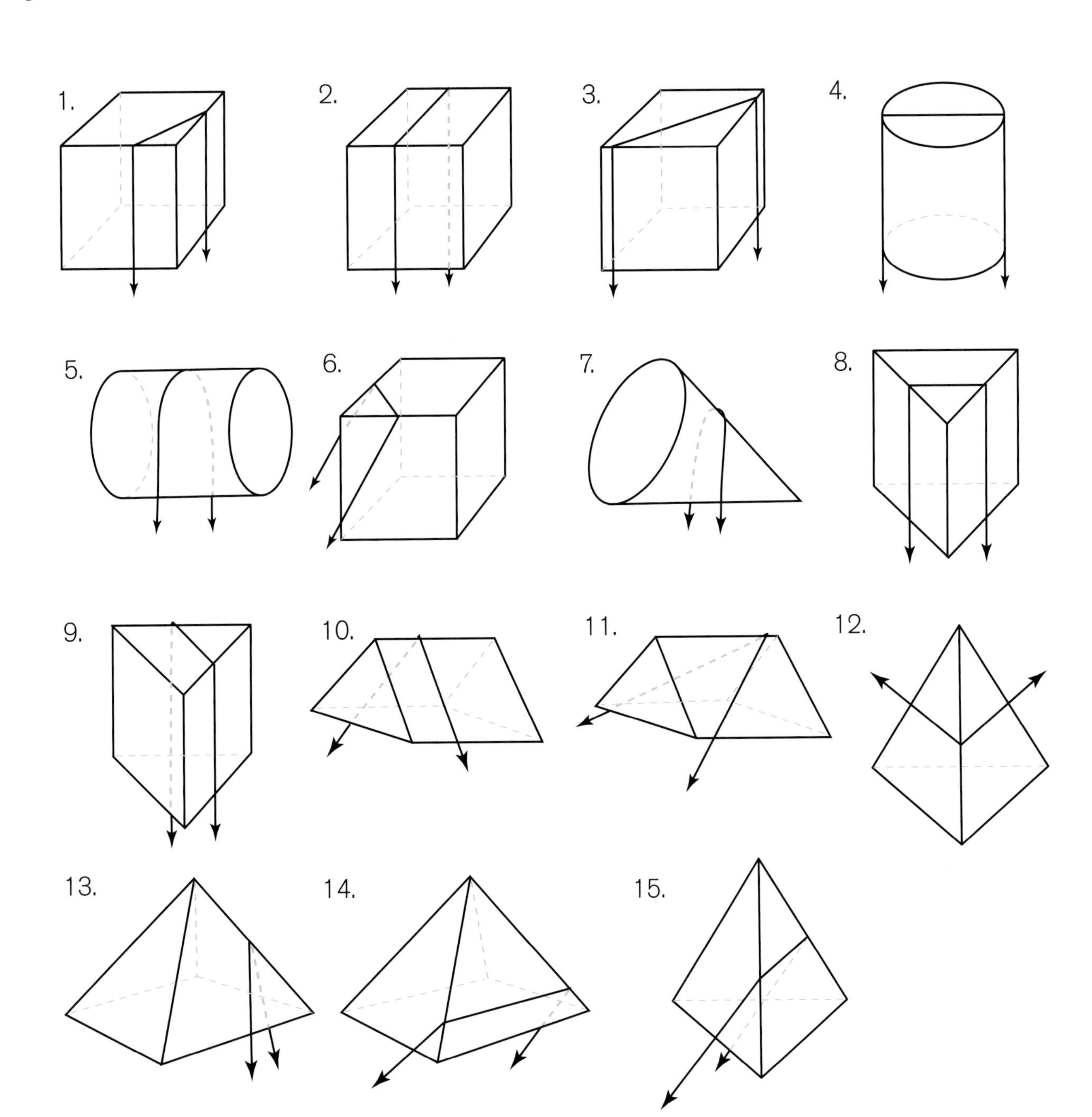

Constructing Three-Dimensional Figures

Name ______________________________

Use the materials your teacher supplies to complete the following tasks:

1. A picture of a solid (a triangular prism) is shown.
   (*a*) Use your materials to build the solid.
   (*b*) Count the number of vertices, edges, and faces.
   (*c*) Record the data in the table at the end of the activity sheet.
   (*d*) Open your solid to form a net. Draw the net on a separate piece of paper.
   (*e*) Draw a different net that can be used to make the solid.
   (*f*) Describe the relationship between the nets you have drawn and the solid.

____

____

____

____

2. A picture of another solid (a hexagonal pyramid) is shown.
   (*a*) Use your materials to build the solid.
   (*b*) Count the number of vertices, edges, and faces.
   (*c*) Record the data in the table below.
   (*d*) Open your solid to form a net. Draw the net on a separate piece of paper.
   (*e*) Draw a different net that can be used to make the solid.
   (*f*) Describe the relationship between the nets you have drawn and the solid.

____

____

____

____

Name ______________________________

3. Build, sketch, and examine some solids:

   (*a*) Build a solid using a total of eight squares and triangles.

   (*b*) Sketch your solid on a separate piece of paper.

   (*c*) Build a different solid using only eight squares and triangles.

   (*d*) On a separate piece of paper, sketch your solid.

   (*e*) How many different solids can you build using only eight Polydron pieces? ______________
   Sketch each solid on a separate piece of paper.

   (*f*) Count the number of vertices, edges, and faces on the solids you built. Record your data in the table. What patterns do you observe in the data? ______________________________

   ____

   ____

Solid	Vertices	Edges	Faces
1			
2			
3			
4			
5			
6			
7			
8			

Indirect Measurement

Name ____________________

1. To find the width of the natural rock formation shown at the right, a surveyor located a point (*A*) from which she could sight the edges of the rock. From *A*, she drew rays through both edges of the rock (call them points *B* and *C*) and beyond. Next she placed a marker, *E*, on ray *AC* 20 feet beyond point *C*. Then she placed a second marker, *D*, on ray *AB* so that the ratios *AC*/*AE* and *AB*/*AD* were equal. Finally, she constructed segment *DE*. Show that *DE* is parallel to *BC*, and find the distance through the rock formation.

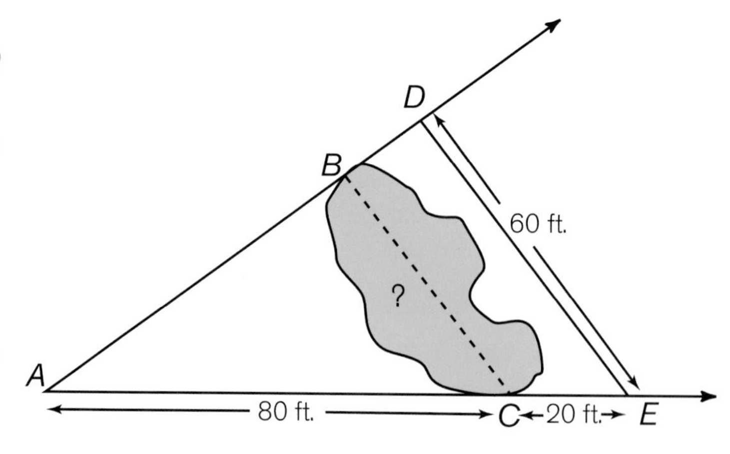

2. Similar triangles were contructed as a way to find the distance across the pond shown at the right. If $\overline{AD}$ is 30 meters, $\overline{BD}$ is 10 meters, and $\overline{BC}$ is 24 meters, find $\overline{DE}$, the distance across the pond.

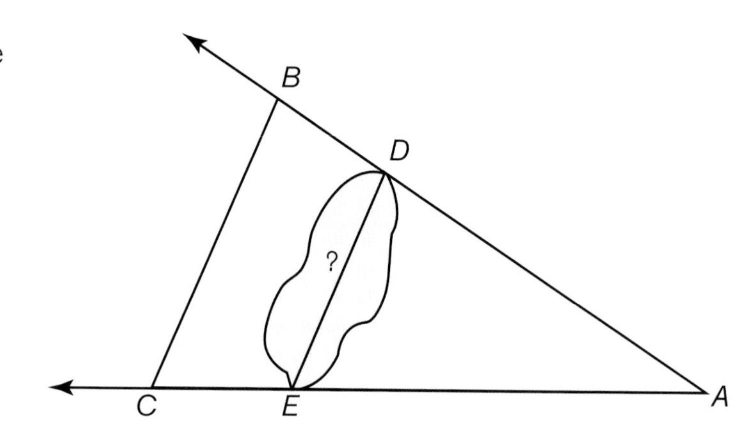

3. Describe a method for estimating the height of the large tree if the small tree is 4 feet tall.

____

____

____

____

____

____

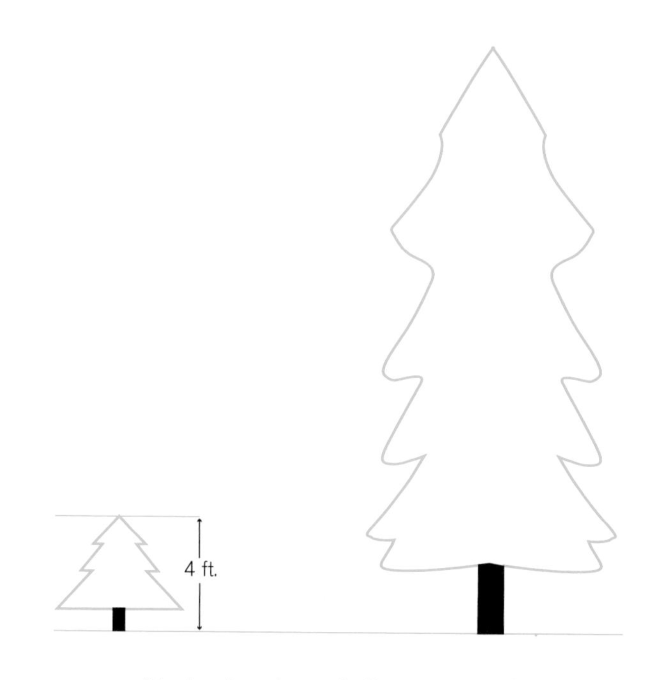

The Race

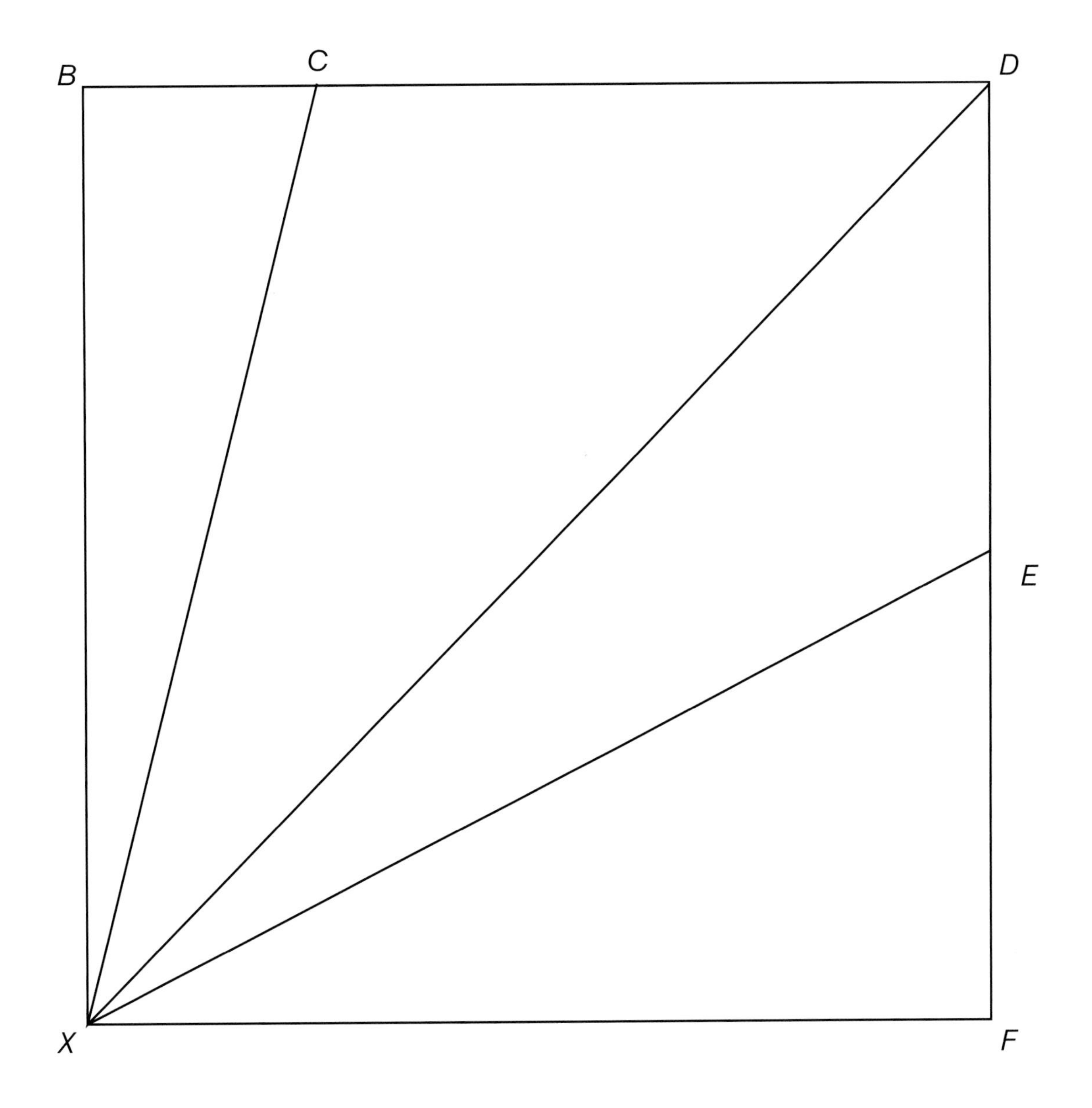

Solutions for the Blackline Masters

Solutions for "Geodee's Sorting Scheme"

1–3. Answers will vary. Shape O could be placed in either category, depending on how the categories are defined. If category 2 contains regular polygons, the shape goes in category 1. If category 2 consists of polygons with congruent sides, shape O could be placed in it.

Solutions for "Exploring Triangles"

2. (*a*) As angle *BCA* moves closer to segment *AB*, the measure of angle *BCA* becomes greater; (*b*) the length of side *AB* remains the same, whereas the lengths of sides *AC* and *BC* become shorter.

3. (*a*) The measure of angle *BCA* becomes 180°; (*b*) the sides of the triangle become coincident with side *AB*, and the sum of their lengths equals the length of *AB*.

4. (*a*) Angles ABC and BAC have equal measure, and the sum of all three measures is 180°; (*b*) the angles *BAC* and *BCA* would be equal, and if all three sides were made equal, all three angles would be equal.

5. Answers will vary. The following are some possible responses:
   - In a triangle, the angle with the largest measure is opposite the longest side.
   - In a triangle, if two sides have equal measures, the angles opposite those two sides also have equal measures.
   - In a triangle, if the three sides have equal measures, then the three angles have equal measures.

Solutions for "Exploring Similar Figures"

Students' descriptions of their methods should focus on keeping the angles congruent and the sides proportional. The students could measure the angles and the lengths of the sides to help them come to some conclusions about the figures. The students should realize that enlargements produce similar figures (figures that have congruent angles—that is, they stay the same—and sides that are proportional). The lengths of the sides of the new figure are 1.5 times the lengths of the sides of the original figure, so the scale factor is 3/2. The enlargement appears at the right.

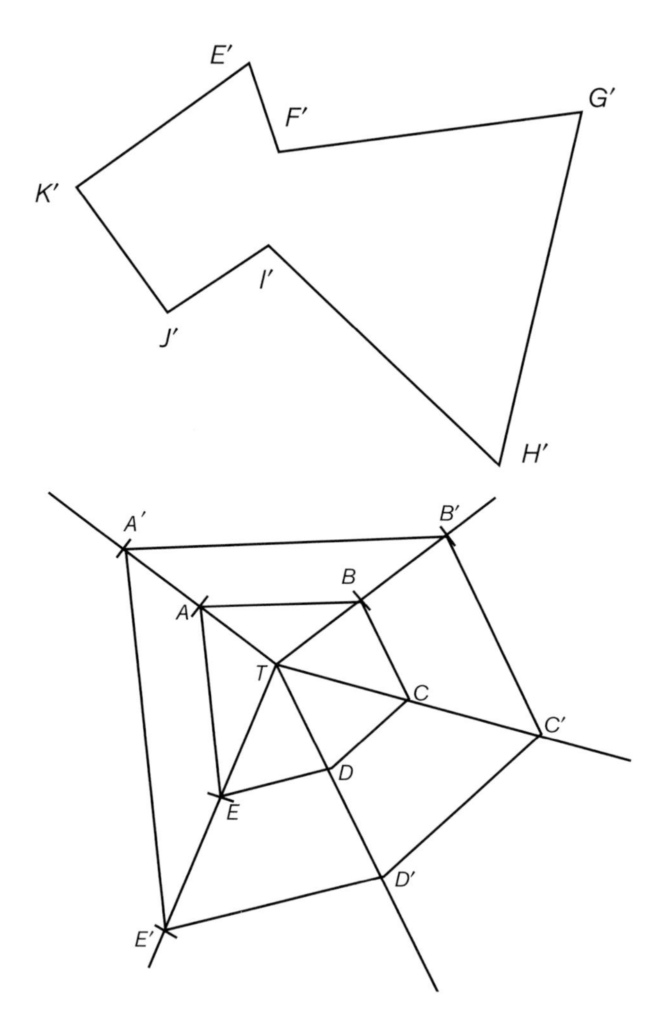

Solutions for "Dilating Figures"

1. Sheri is using point *T* as the center of dilation. From *T*, she draws rays extending through the vertices of *ABCDE* to obtain the points for the vertices of the dilated pentagon. She places the vertices at the points whose distances from *T* are twice the length of the segments from *T* to the corresponding vertices of the original pentagon.

2. See the figure at the right.

3. Many answers are possible. For instance, the students may notice that the corresponding angles of the original and

the dilated figures are congruent and that the corresponding sides are proportional. They also may conjecture that the corresponding sides of the two pentagons are parallel.

Solutions for "Using Venn Diagrams to Reason about Shapes"

1–4. Answers will vary.

Solutions for "Midsegments of Triangles"

1–3. Answers will vary.

4. Answers will vary; many students may notice that the lengths of the midsegments are half the lengths of the sides that are parallel to the midsegments and that the angles of the triangles formed by the midsegments are congruent to the corresponding angles of the original triangles.

Solutions for "Reasoning about the Pythagorean Relationship"

2–3.

Triangle Number	Length of Side *a* (Units)	Length of Side *b* (Units)	Length of Side *c* (Units)	Area of Square with Length *a* (Square Units)	Area of Square with Length *b* (Square Units)	Area of Square with Length *c* (Square Units)
1	3	4	5	9	16	25
2	5	12	13	25	144	169
3						
4						

4. The students may discover that the area of the square with length *c* equals the combined area of the squares with lengths *a* and *b*.

Solutions for "Constructing Geometric Figures in Coordinate Space"

1–14. Answers will vary.

Solutions for "Exploring Lines, Midpoints, and Triangles Using Coordinate Geometry"

Graphs, Coordinates, and Lines

2. Point (8, 9) is out of place.
3. The coordinates of the other five points lie on the same line, $y = x + 2$.
5. The five points in the graph lie on the same vertical line.
6. All five points have the same *x*-coordinate.

Graphs, Coordinates, and Midpoints

2. The coordinates of point *M* are (3, 4).
3. The *x*-coordinate of point *M* is the sum of the x-coordinates of points *A* and *B* divided by 2; the *y*-coordinate of point *M* is the sum of the *y*-coordinates of points *A* and *B* divided by 2.
4. The coordinates of point *N* are (9, 8).

5. Yes, they are related in the same way. One observation may be that each midpoint is on the same line as the two given points. Another possible observation is that both points are found in the same way (see answer 3, above).

Graphs, Coordinates, and Triangles

2. The students should notice that the triangles appear congruent. The congruence of the triangles should become more apparent as the students answer questions 3 and 4.
3. The *y*-coordinates of the vertices of triangle 2 are four more than the *y*-coordinates of the corresponding vertices of triangle 1.
4. The *x*-coordinates of the vertices of triangle 3 are seven more than the *x*-coordinates of the vertices of triangle 1; the *y*-coordinates of the vertices of triangle 3 are one less than the *y*-coordinates of the vertices of triangle 1.
6. Students may notice that the triangles are different sizes but have the same shape; that is, they are similar.
7. The coordinates of the vertices of triangle 1 can be multiplied by 2 to get the coordinates of the vertices of triangle 2, $(2x_1, 2y_1)$.
8. The coordinates of the vertices of triangle 1 can be multiplied by 1/2 to get the coordinates of the vertices of triangle 3, $((1/2)x_1, (1/2)y_1)$.

Solutions for "Similarity and the Coordinate Plane"

1. The points *A*′, *B*′, *C*′, *D*′, *E*′, *F*′, *G*′, and *H*′ lie on rays *OA, OB, OC, OD, OE, OF, OG,* and *OH,* respectively. See figure 2.4.
2. The distance of *OA*′ is twice that of *OA,* and the distance of *OB*′ is twice that of *OB*. Multiplying the coordinates by 2 doubles the distance from the origin.
3. The two triangles and the two pentagons have the same shape.
4. The distances from the origin to the vertices of the resulting shapes would be three times, four times, and one-fifth the distance, respectively, from the origin to the corresponding vertices of the original figures.
5. Answers will vary.

Solutions for "Exploring the Slopes of Parallel and Perpendicular Lines"

1. The slopes of $\overline{JK}$ and $\overline{ML}$ are both 3/2, the slope of $\overline{JM}$ is −2, and the slope of $\overline{KL}$ is 1/3; the slope of $\overline{AB}$ is −1/2, of $\overline{BC}$ is 2, and of $\overline{AC}$ is 1; the slope of $\overline{NQ}$ is −1, of $\overline{NO}$ is 5/4, of $\overline{OP}$ is −1, and of $\overline{QP}$ is 5/4.
2. (*a*) The slopes of $\overline{JK}$ and $\overline{ML}$ are the same, 3/2; the slopes of $\overline{NQ}$ and $\overline{OP}$ are both −1; the slopes of $\overline{NO}$ and $\overline{QP}$ are both 5/4.

   (*b*) The sides with the same slope are parallel.

   (*c*) *NOPQ* is a parallelogram, a quadrilateral with two pairs of parallel sides; *JKLM* is a trapezoid, a quadrilateral with exactly one pair of parallel sides.

   (*d*) The slopes of sides *AB* and *BC* are negative reciprocals (−1/2 and 2). $\angle ABC = 90°$.
3. The slopes of parallel lines are the same, and the slopes of perpendicular lines are negative reciprocals.

Solutions to "Reflection of Images"

For each of the figures, the image of the figure reflected over the *y*-axis appears in green and the image of the figure reflected over the *x*-axis appears in gray.

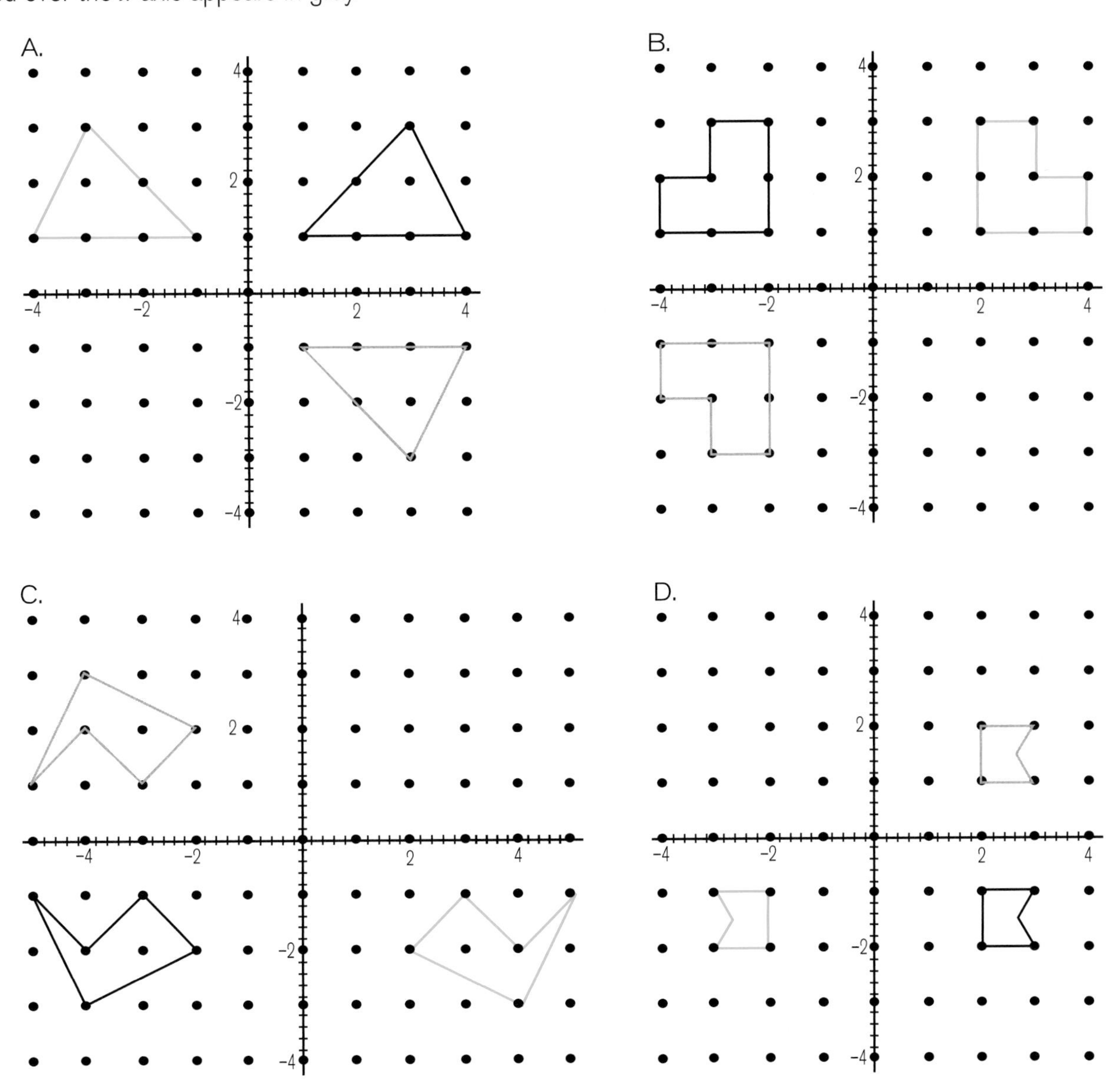

Solutions for "Using Scale Factors"

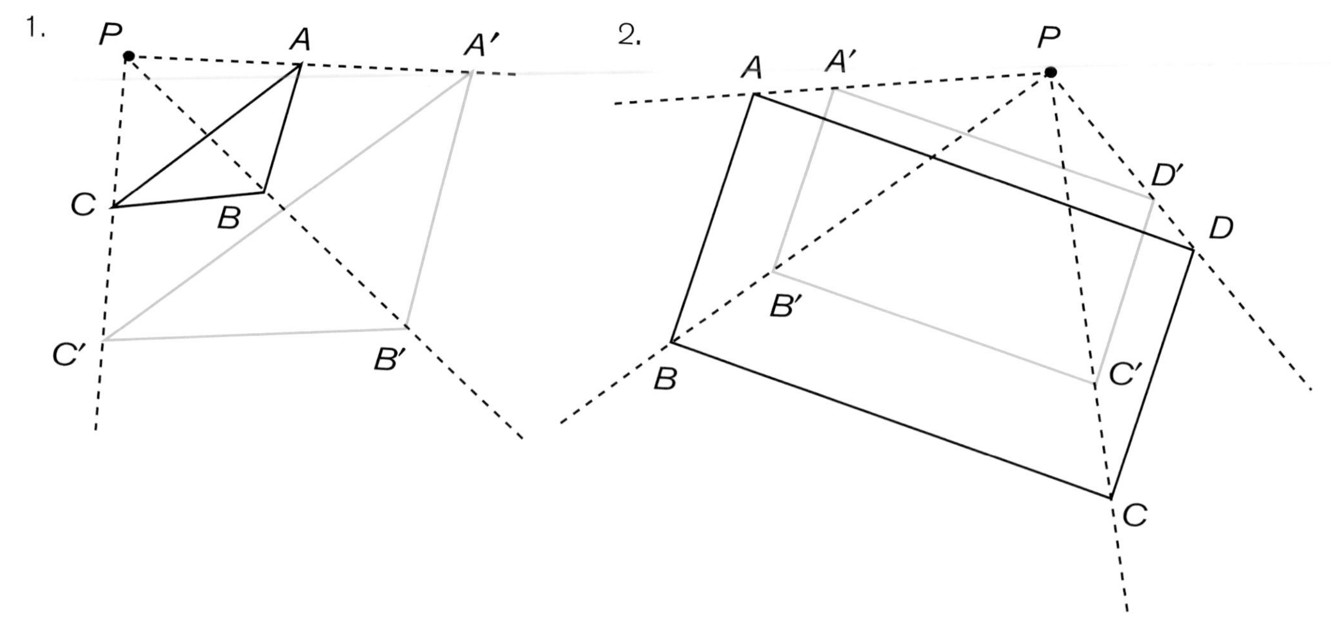

Solutions for "Explorations with Lines of Symmetry"

1. The dashed lines in figures A and D are lines of symmetry. The dashed lines in figures B and C are not lines of symmetry. Students' explanations for their responses might include that the parts of the figure on either side of the dashed line are (or are not) mirror images of each other.
2. Answers will vary.

Solutions for "Rotational Symmetry and Regular Polygons"

5.

Rotational Symmetry in Regular Polygons

Number of sides	3	4	5	6	7	8	. . .	n
Number of rotational symmetries	3	4	5	6	7	8	. . .	n
List of rotational symmetries	120° 240° 360°	90° 180° 270° 360°	72° 144° 216° 288° 360°	60° 120° 180° 240° 300° 360°	~51.4° ~102.8° ~154.2° ~205.6° ~257.0° ~308.4° 360°	45° 90° 135° 180° 225° 270° 315° 360°	. . .	$\frac{360^\circ}{n}$ $2\left(\frac{360^\circ}{n}\right)$ $3\left(\frac{360^\circ}{n}\right)$ $\vdots$ $n\left(\frac{360^\circ}{n}\right)$

Conjecture: The number of sides of a regular polygon is equal to the number of rotational symmetries of the polygon.

Solutions for "Logos and Geometric Properties"

1–2. Answers will vary.

Solutions for "Isometric Explorations"

1. e; 2. b; 3. All are views (a. right, b. front, c. top, d. back, e. left); 4. d; 5. b; 6. See drawing.

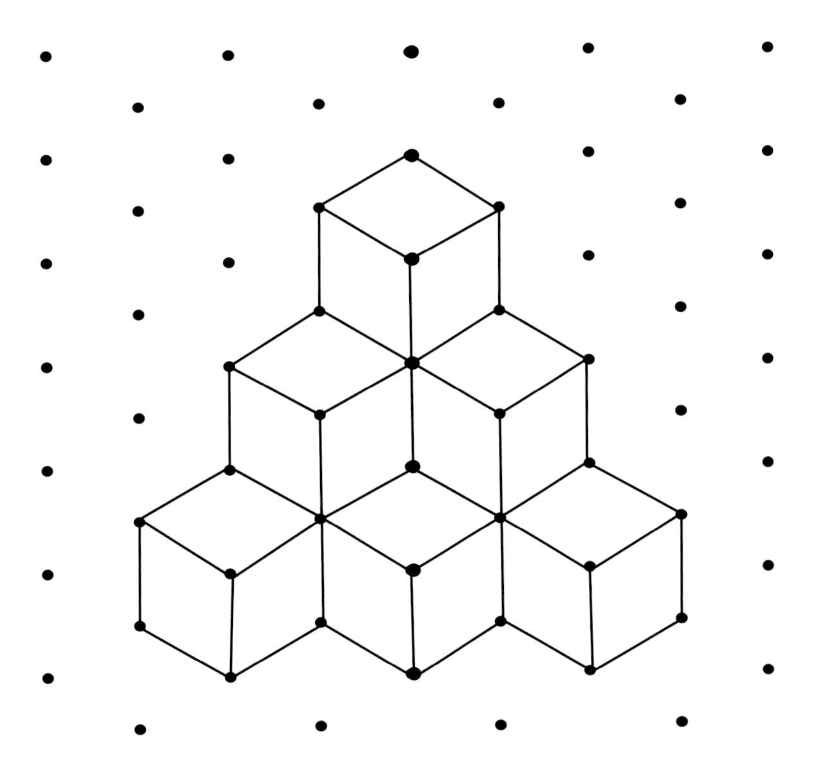

Solutions for "Cross Sections of Three-Dimensional Shapes"

The following answers represent the general shapes of the cross sections. In some cases, the ratios of the sides may vary depending on the placement of the "slice."

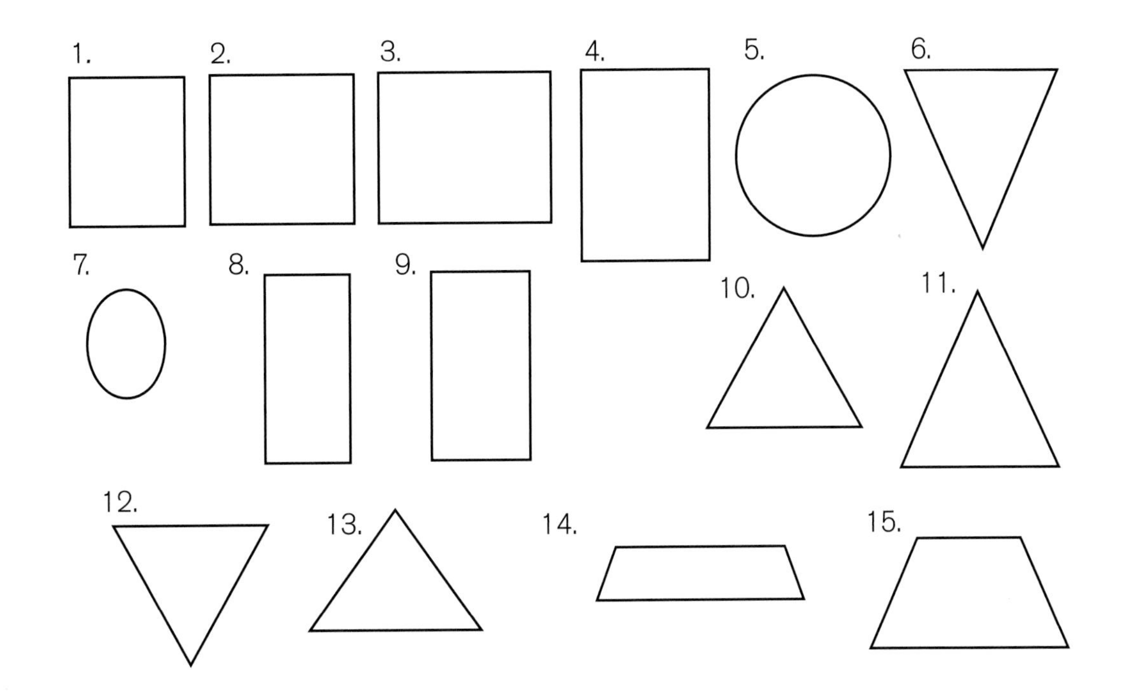

Solution to "I Took a Trip on a Train"

The correct order for the photographs is 1, 4, 2, 3.

Solutions for "Constructing Three-Dimensional Figures"

1 and 2. See the table for the answers to 1 (c) and 2 (c). The other answers will vary.
3. Answers will vary.

Solid	Vertices	Edges	Faces
1	6	9	5
2	7	12	7
3			
4			

Solutions for "Indirect Measurement"

1. $\overline{BC} = 48$ feet.
2. $\overline{DE} = 18$ meters.
3. For this problem, students could consider the shadows cast by the trees. For example, they might assume that the shadow cast by the large tree is twenty-five feet and the shadow cast by the small tree is five feet. By proportional reasoning, then, they would discover that the large tree is five times the height of the small tree, or twenty feet tall. To see how similar triangles are involved, consider the following sketch.

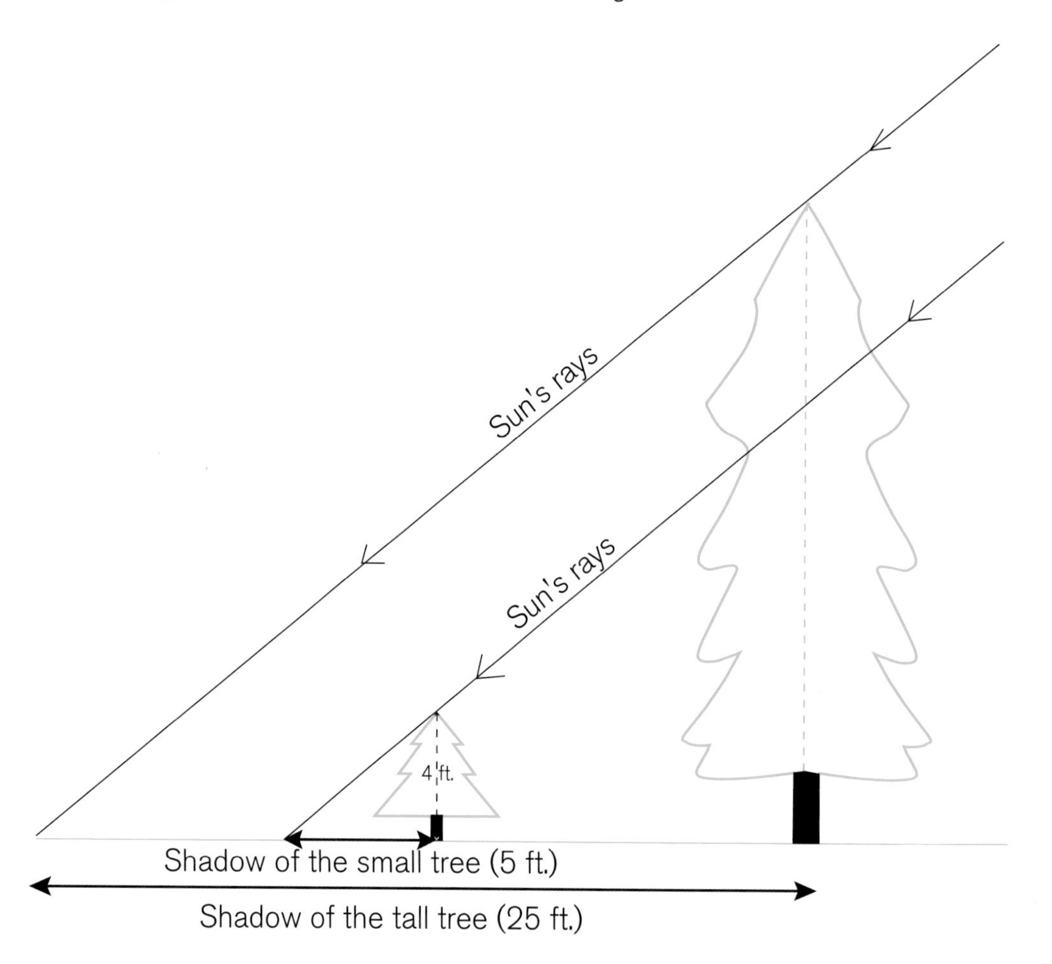

References

Abbott, Edwin A. *Flatland: A Romance of Many Dimensions.* Boston: Little & Brown, 1937.

Bippert, Judy. "Cube Challenge." *Mathematics Teacher* 86 (May 1993): 386–90, 395–98.

Cathcart, George W., Yvonne M. Pothier, James H. Vance, and Nadine S. Bezuk. *Learning Mathematics in Elementary and Middle Schools.* Upper Saddle River, N.J.: Prentice Hall, 2000.

Clements, Douglas H., and Michael T. Battista. "Geometry and Spatial Reasoning." In *Handbook of Research on Mathematics Teaching and Learning,* edited by Douglas A. Grouws, pp. 420–64. New York: Macmillan Publishing Co. and National Council of Teachers of Mathematics, 1992.

Curriculum Research and Development Group (CRDG). "Reshaping Mathematics" (draft). University of Hawaii at Manoa, Honolulu, Hawaii, n.d. Photocopied.

Curriculum Research and Development Group, Geometry Learning Project. 'Teacher Materials" (draft). University of Hawaii at Manoa, Honolulu, Hawaii, n.d. Photocopied.

Friedlander, Alex, and Glenda Lappan. "Similarity: Investigations at the Middle Grades Level." In *Learning and Teaching Geometry, K–12,* 1987 Yearbook of the National Council of Teachers of Mathematics, edited by Mary Montgomery Lindquist, pp. 136–43. Reston, Va.: National Council of Teachers of Mathematics, 1987.

Geddes, Dorothy, and Irene Fortunato. "Geometry: Research and Classroom Activities." In *Research Ideas for the Classroom: Middle Grades Mathematics,* edited by Douglas T. Owens, pp. 199–222. New York: Macmillan Publishing Co., 1993.

Germain-McCarthy, Yvelyne. "The Decorative Ornamental Ironwork of New Orleans: Connections to Geometry and Haiti." *Mathematics Teaching in the Middle School* 4 (April 1999): 430–36.

Keiser, Jane M. "The Role of Definition." *Mathematics Teaching in the Middle School* 5 (April 2000): 506–11.

Kelley, Paul. "Build a Sierpinski Pyramid." *Mathematics Teacher* 92 (May 1999): 384–86.

Lappan, Glenda, James T. Fey, William M. Fitzgerald, Susan N. Friel, and Elizabeth Difanis Phillips. *Ruins of Montarek: Spatial Visualization.* Parsippany, N.J.: Dale Seymour Publications, 1996.

National Council of Teachers of Mathematics (NCTM). *Principles and Standards for School Mathematics.* Reston, Va.: NCTM, 2000.

O'Daffer, Phares G., and Stanley R. Clemens. *Geometry: An Investigative Approach.* 2nd ed. New York: Addison-Wesley Publishing Co., 1992.

PBS. "PBS Mathline." Grades 6–8 Activity 3: Dynamic Geometry. PBS TeacherSource. 2001. www.pbs.org/teachersource/mathline/concepts /technology/activity3.shtm (9 September 2001)

Pugalee, David K. "Using Communication to Develop Students' Mathematical Literacy." *Mathematics Teaching in the Middle School* 6 (January 2001): 296–99.

Reconceptualizing Mathematics Project. "Shapes and Measurement Module" (draft, pp. 198–201). Center for Research in Mathematics Education, San Diego State University, San Diego, Calif., n.d. Photodcopied.

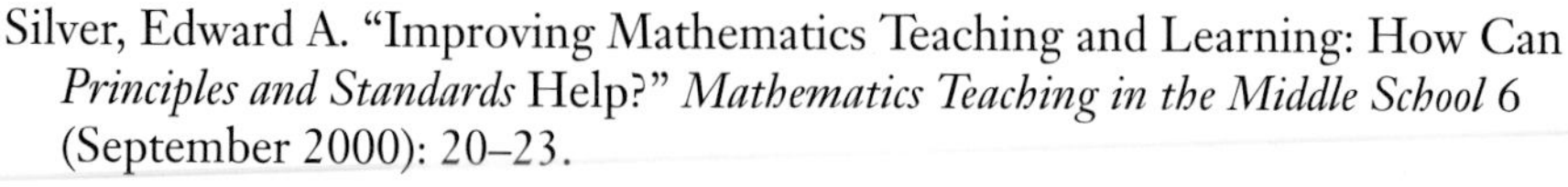
Silver, Edward A. "Improving Mathematics Teaching and Learning: How Can *Principles and Standards* Help?" *Mathematics Teaching in the Middle School* 6 (September 2000): 20–23.

Slavit, David. "Above and beyond AAA: The Similarity and Congruence of Polygons." *Mathematics Teaching in the Middle School* 3 (January 1998): 276–80.

Stein, Mary K., Margaret S. Smith, Marjorie A. Hennigsen, and Edward A. Silver. *Implementing Standards-Based Mathematics Instruction.* New York: Teachers College Press, 2000.

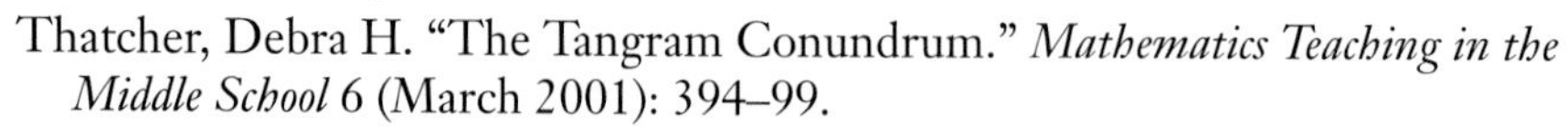
Thatcher, Debra H. "The Tangram Conundrum." *Mathematics Teaching in the Middle School* 6 (March 2001): 394–99.

Tillotson, Marian L. "The Effect of Instruction in Spatial Visualization on Spatial Abilities and Mathematics Problem Solving." Ph.D. diss., University of Florida, 1984.

Van de Walle, John A. *Elementary and Middle School Mathematics.* New York: Longman, 1998.

Westegaard, Susanne K. "Stitching Quilts into Coordinate Geometry." *Mathematics Teacher* 91 (October 1998): 587–95.

Winter, Mary Jean, Glenda Lappan, Elizabeth Phillips, and William Fitzgerald. *Middle Grades Mathematics Project: Spatial Visualization.* Menlo Park, Calif.: Addison-Wesley Publishing Co., 1986.

Suggested Reading

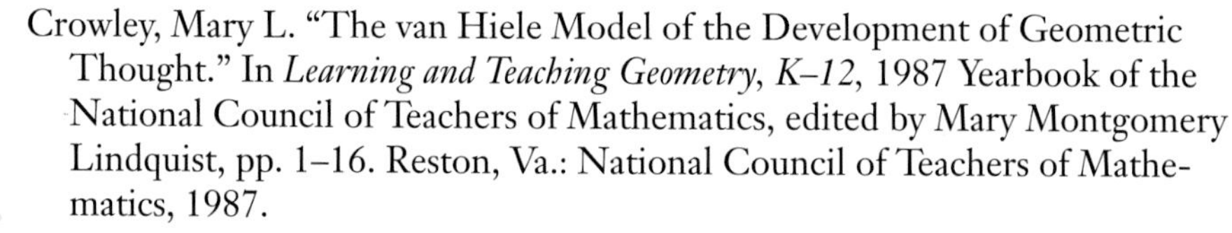
Crowley, Mary L. "The van Hiele Model of the Development of Geometric Thought." In *Learning and Teaching Geometry, K–12,* 1987 Yearbook of the National Council of Teachers of Mathematics, edited by Mary Montgomery Lindquist, pp. 1–16. Reston, Va.: National Council of Teachers of Mathematics, 1987.

Malloy, Carol E. "Perimeter and Area through the van Hiele Model." *Mathematics Teaching in the Middle School* 5 (October 1999): 87–90.

Pohl, Victoria. "Visualizing Three Dimensions by Constructing Polyhedra." In *Learning and Teaching Geometry, K–12,* 1987 Yearbook of the National Council of Teachers of Mathematics, edited by Mary Montgomery Lindquist, pp. 144–54. Reston, Va.: National Council of Teachers of Mathematics, 1987.